American Environmentalism

American Environmentalism
The U.S. Environmental Movement, 1970–1990

Edited by
Riley E. Dunlap and Angela G. Mertig
Washington State University
Pullman, Washington

USA	Publishing Office:	Taylor & Francis 1101 Vermont Ave., NW, Suite 200 Washington, DC 20005-3521
	Sales Office:	Taylor & Francis Inc. 1900 Frost Road, Bristol, PA 19007-1598
UK		Taylor & Francis, Ltd. 1 Gunpowder Square London EC4A 3DE

The papers in this book were originally published in the journal *Society and Natural Resources,* © 1990, 1991 Taylor & Francis.

American Environmentalism: The U.S. Environmental Movement, 1970–1990

5 6 7 8 9 0   B R B R   9 8 7 6

This book was set in Times Roman by Hemisphere Publishing Corporation. The editors were Marly Davidson and Carolyn V. Ormes, the production supervisor was Peggy M. Rote, and the typesetter was Ana Alvandian. Cover design by Michelle Fleitz. Printing and binding by Braun-Brumfield.

A CIP catalog record for this book is available from the British Library.

∞ The paper in this publication meets the requirements of the ANSI Standard Z39.48-1984 (Permanence of Paper).

Library of Congress Cataloging in Publication Data

American environmentalism: the U.S. environmental movement, 1970–1990
/ edited by Riley E. Dunlap and Angela G. Mertig.
p. cm.
"Originally published in the journal, Society and natural resources"—T.p. verso.
Includes bibliographical references and index.

1. Environmental policy—United States—Citizen participation—History—20th century. 2. Green movement—United States—History—20th century. I. Dunlap, Riley E. II. Mertig, Angela G.
HC110.E5A648 1992
363.7′057′0973—dc20 92-7035
ISBN 0-8448-1730-9 CIP

This book is dedicated to our parents:

Jean Anderson and the memory of Riley W. Dunlap, Jr.
and
Ted and Barbara Mertig

Contents

Preface *xi*

Chapter 1 **The Evolution of the U.S. Environmental Movement from 1970 to 1990: An Overview** **1**
Riley E. Dunlap and Angela G. Mertig

The Emergence of the U.S. Environmental Movement: A Brief History 1
The Natural History of Social Movements 3
The Stages of Modern Environmentalism 4
Increased Diversity: Strength or Weakness? 5
Conclusion: When Is a Movement Successful? 8
Notes 8
References 9

Chapter 2 **Twenty Years of Environmental Mobilization: Trends Among National Environmental Organizations** **11**
Robert Cameron Mitchell, Angela G. Mertig, and Riley E. Dunlap

Evolution of the National Environmental Organizations 12
Activities of National Organizations 19
What Success Has Meant for the National Organizations 23
Conclusion 24
Notes 24
References 25

Chapter 3 **Not in Our Backyards: The Grassroots Environmental Movement** **27**
Nicholas Freudenberg and Carol Steinsapir

Description of the Grassroots Environmental Movement 28
Shared Perspectives Among Grassroots Environmentalists 31
Comparison with the National Environmental Organizations 32
Accomplishments of the Grassroots Environmental Movement 33
The Future: From NIMBY to NIABY 35
References 35

Chapter 4 **The Quest for Environmental Equity: Mobilizing the African-American Community for Social Change** **39**
Robert D. Bullard and Beverly H. Wright

The Environmental Justice Movement 40
The Attraction of Environmental Justice 41
Case Studies from the Southern United States 43
Environmental Conflict Resolution 45
Conclusion 47
Acknowledgment 48
References 48

Chapter 5 **Deep Ecology and Radical Environmentalism** **51**
Bill Devall

Deep Ecology 52
Deep Ecology and Green Politics 54
Deep Ecology and Radical Environmentalism 55
Radical Environmental Movements 57
Trends in Radical Environmentalism 59
Notes 61
References 61

Chapter 6 **Globalizing Environmentalism: Threshold of a New Phase in International Relations** **63**
Lynton K. Caldwell

Legitimizing Global Environmentalism: The Legacy of Stockholm 64
Globalizing Popular Concern 67
Implementing Global Environmentalism: Political and Institutional Challenges 72
Notes 74
References 75

Chapter 7 **Twenty Years of Change in the Environmental Movement: An Insider's View** **77**
Michael McCloskey

Ideology 77
Issues 80
Strategies 81
Tactics 82
How the National Organizations Have Changed 83
Challenges Ahead 85
Notes 88
References 88

Chapter 8 **Trends in Public Opinion Toward Environmental Issues: 1965–1990** **89**
Riley E. Dunlap

The Natural Decline Model 90
Trends in Public Concern for Environmental Quality 91
Public Perceptions of Ecological Problems in the 1980s 106
Summary and Implications 112
Notes 114
References 115

Index *117*

About the Editors *123*

Preface

The second half of the twentieth century has seen the emergence of numerous social movements in the United States. Most of these movements have faded away with little discernible impact, but history will surely record the environmental movement as among the few that significantly changed our society. The twentieth anniversary of Earth Day in 1990 not only signified the success of the environmental movement, but provided a fitting occasion for taking stock of it—especially how the movement has evolved since bursting upon the national scene with the celebration of the first Earth Day on April 22, 1970.

Taking stock of the movement was the purpose of a symposium on "Two Decades of Environmentalism" held at the 1990 meeting of the American Association for the Advancement of Science. This symposium brought together several leading analysts of environmentalism. The goal of the symposium was to describe the evolution of the movement over the past two decades and to assess its current status, particularly its growing diversity. Revisions of the symposium papers were subsequently published in late 1991 as a special issue of *Society and Natural Resources,* and the present volume is a slight expansion of that issue designed for easier access by students, scholars, and others interested in environmentalism.

Recognized experts on various aspects of environmentalism were invited to contribute analyses of the major strands of the movement, with the goal of providing a reasonably comprehensive assessment of the past two decades and the current status of the environmental movement. In each case, contributors were asked to provide detailed assessments of their particular aspect of the movement.

Social scientists have examined environmentalism from various theoretical perspectives. Some have portrayed it as one of a myriad of "new social movements" (concerned primarily with noneconomic goals) stimulated by the rise of post-materialist values in the affluent, post–World War II era, and others have emphasized the unique characteristics of environmental problems that engender widespread collective action to solve them or the success with which activists have mobilized a wide range of resources to stimulate such action. Rather than ask the contributors to this volume to frame their analyses in terms of a particular perspective, however, we asked them to provide in-depth descriptions and analyses employing the perspectives and concepts they deemed appropriate and to write for a wide range of readers rather than for academic specialists. We think the result is an accurate and richly detailed depiction of modern environmentalism that will provide readers with a good sense of where the movement has been, where it stands, and where it is headed.

In an effort to place the current movement in historical context, the introductory chapter briefly traces the roots of the environmental movement and then presents a model of the evolution of the typical social movement to provide a baseline from which to evaluate the success of environmentalism. The second chapter, by Mitchell, Mertig, and Dunlap, examines the major national organizations (e.g., the Sierra Club) that constitute the "environmental lobby," the most visible manifestation of the movement, and especially how these organizations have changed over the past two decades.

The next four chapters document the growing diversity within environmentalism by examining the emergence of several major strands or components of the movement. A significant development in the late seventies was the rapid emergence of grassroots organizations concerned primarily with combatting local environmental hazards, epitomized by Love Canal, and this strand of the movement is examined by Freudenberg and Steinsapir in the third chapter. More recent still has been the growing concern over environmental hazards within minority communities. A minority-based environmental justice component of environmentalism became visible in the last half of the eighties, and it is described by Bullard and Wright in chapter 4 (the only chapter that did not stem from the AAAS symposium and special issue of *Society and Natural Resources*). In the fifth chapter, Devall focuses on a highly visible development within environmentalism, the emergence of radical environmentalism such as that embodied by Earth First! A particularly obvious change over the past two decades has been the spread of environmental concerns throughout most of the world, and in chapter 6 Caldwell describes this globalization process and what it has meant for U.S. environmentalists.

The first six chapters were written by academic analysts who are well-versed in the various strands of environmentalism they examine, but to supplement their analyses the seventh chapter provides an overview by a key insider in the movement—the former president and current chairman of the Sierra Club. McCloskey's examination of the past two decades of environmentalism from the perspective of a key participant nicely complements the prior chapters. The eighth and last chapter adds a final dimension to the volume by focusing on public opinion toward environmental issues. Although most of the public remain outsiders vis-à-vis the organized movement, the expression of public support for environmental protection has been a valuable resource for environmental activists, and Dunlap traces long-term trends in this support.

Taken together, we believe that these chapters will provide readers with a good sense of the complex nature of contemporary environmentalism and how it has evolved over the past two decades. We expect that students and scholars interested in environmentalism and in environmental policy and politics, as well as those interested in social problems and social movements, will find the volume of interest. We also hope that scholars, policymakers, activists, and citizens concerned about our society's response to environmental problems will likewise find it of interest.

We thank the contributors to this volume for making its publication possible, especially for their willingness to revise and update their original papers in order to achieve a more integrated, coherent, and timely volume, and Rabel Burdge, Co-Editor of *Society and Natural Resources,* for his support throughout this project. Dunlap also thanks his graduate school advisor, Richard Gale, for collaborating on a 1970 study of University of Oregon "eco-activists" that stimulated a long-term interest in environmentalism as well as for encouragement to pursue these interests. Gale also supervised Devall's dissertation research on the Sierra Club and thus helped launch another career intimately involved with the study of environmentalism. Mertig thanks Eric Godfrey of Ripon College for introducing her to the sociological study of environmental issues and for encouraging her to pursue graduate work in environmental sociology.

Finally, we thank the millions of Americans who have labored on behalf of environmental protection. Their efforts and dedication have kept the environmental movement a vital force in American society for over two decades, a phenomenon that clearly warrants scholarly attention.

Riley E. Dunlap and Angela G. Mertig
Washington State University, Pullman, Washington

The Evolution of the U.S. Environmental Movement from 1970 to 1990: An Overview

RILEY E. DUNLAP
ANGELA G. MERTIG

Departments of Sociology and Rural Sociology
Washington State University
Pullman, WA 99164-4006
USA

Abstract *The twentieth anniversary of Earth Day in 1990 indicated that environmentalism has become an enduring and successful social movement and thus has avoided the rapid demise that many movements experience. However, the U.S. environmental movement has changed considerably since 1970, most notably by becoming a much more diverse movement. The contributions to this volume describe the major strands of environmentalism that have developed over the past two decades and, in the process, provide an overview of the evolution of the movement.*

Keywords Environmental movement, environmentalism, interest groups, social movements, social problems.

The U.S. environmental movement has proven to be exceptionally successful and enduring, as demonstrated by the twentieth anniversary of Earth Day on 22 April 1990. A national poll conducted a week earlier found nearly two-thirds (63%) of Americans to be in support of Earth Day and only a tiny fraction (3%) opposed to it (Hart/Teeter, 1990). It is not surprising, therefore, that Earth Day 1990 turned out to be an enormously successful event, both in the United States and worldwide, leading to the claim that "it united more people concerned about a single cause than any other global event in history" (Cahn and Cahn, 1990, p. 17).

The success of Earth Day 1990 indicates that the environmental movement is not only alive and well after two decades but that it may be stronger than ever. Few social movements achieve such widespread acceptance, and fewer still are able to celebrate a twentieth anniversary. Why has environmentalism been able to avoid the fate of most short-lived movements, and how has it changed since the first Earth Day? Such questions are the focus of this volume, which is intended to provide an overview of the evolution of U.S. environmentalism during the past two decades.[1]

The Emergence of the U.S. Environmental Movement: A Brief History

The organizational and ideological roots of contemporary environmentalism are commonly traced to the progressive conservation movement that emerged in the late nineteenth century in reaction to reckless exploitation of our nation's natural resources. Early conservationists led by Gifford Pinchot (with support from Theodore Roosevelt) emphasized the wise management of natural resources for continued human use, but a few individuals such as John Muir argued for the preservation of nature for its own sake. Although the two factions eventually came into conflict, their joint efforts led to legislation establishing early national parks and agencies such as the U.S. Forest Service. They also spawned conservation organizations such as the Sierra Club and the National Audubon Society.[2]

World War I deflected the nation's attention from conservation (O'Riordan, 1971), but after the war the United States was confronted by massive environmental calamities such as flooding and the Dust Bowl, as well as by the Great Depression. A second wave of conservationism arose during the Franklin Roosevelt administration and emphasized the mitigation of resource problems (e.g., flood control and soil conservation) as well as the development of resources (e.g., energy through the Tennessee Valley Authority) to stimulate economic recovery. Although these efforts were sidetracked by World War II, the 1950s saw a third wave of conservationism. More emphasis was placed on preservation of areas of natural beauty and wilderness for public enjoyment in this era, spearheaded by older organizations such as the Sierra Club. Widely publicized efforts to save the Grand Canyon and Dinosaur National Monument provided these organizations with considerable momentum. The resulting "wilderness movement" (McCloskey, 1972) was accompanied by continued concern about the future availability of natural resources as well as by growing concerns about overpopulation and air and water quality (Paehlke, 1989).

These old and new issues began to coalesce in the 1960s and gradually evolved into environmental concerns. Epitomized by Rachel Carsons' (1962) analysis of the subtle and wide-ranging impacts of pesticides on the natural environment and human beings in *Silent spring*, these newer concerns were much broader than those of conservation. Environmental problems tended to (a) be more complex in origin, often stemming from new technologies; (b) have delayed, complex, and difficult-to-detect effects; and (c) have consequences for human health and well-being as well as for the natural environment (Mitchell, 1989). Encompassing both pollution and loss of recreational and aesthetic resources, such problems were increasingly viewed as threats to our quality of life (Hays, 1987).

Thus, by the late 1960s the third wave of conservationism had evolved into modern environmentalism, with the transformation formalized by the national celebration of Earth Day 1970.[3] With its reported 20 million participants (Dunlap and Gale, 1972), Earth Day not only marked the replacement of conservation with the full panoply of environmental issues but it mobilized (albeit temporarily) a far broader base of support than had any of the prior waves of conservationism (Mitchell, 1989). What accounted for this transformation?

Analyses of the emergence of environmentalism (e.g., Hays, 1987) have emphasized one or more of the following:

(1) The 1960s had given rise to an activist culture that encouraged people, especially youths, to take direct action to solve society's ills.[4]
(2) Scientific knowledge about environmental problems such as smog began to grow, as did media coverage of such problems and major accidents such as the 1969 Santa Barbara oil spill.
(3) A rapid increase in outdoor recreation brought many people into direct contact

with environmental degradation and heightened their commitment to preservation.

(4) Perhaps most fundamentally, tremendous post–World War II economic growth created widespread affluence, eventually lowering concern with materialism and generating concern over the quality of life.

(5) Many of the existing conservation organizations broadened their focus to encompass a wide range of environmental issues and attracted substantial support from foundations, enabling them to mobilize increased support for environmental causes. In the process they transformed themselves into environmental organizations (Mitchell, 1989).

Although its origins are undoubtedly complex, the environmental movement had clearly "arrived" by 1970, as shown by the tremendous growth in the size of the conservation-era organizations, the development of newer organizations, and in widespread public support. The emergence of the environmental movement in the 1960s and early 1970s was accompanied by creation of new federal agencies, such as the Environmental Protection Agency and the Council on Environmental Quality, and by legislation aimed at combating air and water pollution and requiring "environmental impact statements." In short, by the early 1970s, society had accepted environmentalists' view of environmental quality as a social problem.

The Natural History of Social Movements

The importance of social movements in generating "social problems" has received considerable attention from sociologists, and some have argued that the two are analytically inseparable (Mauss, 1975, 1989). In this view, proponents of problems coalesce into one or more social movement organizations that attempt to mobilize others to work to ameliorate the problematic conditions (environmental degradation, racism, sexism, etc.). These organizations must obtain the support of the media, funding sources, the public, and ultimately of policymakers. Although many such "social-problem movements" fail to get off the ground, some are successful in mobilizing enough support to generate societal action aimed at solving the problem.

Solutions generally take the form of new government regulations and agencies, which signify the institutionalization of the movement. In the process of achieving such success a movement typically loses momentum: Its organizations evolve into formalized interest groups staffed by activists-turned-bureaucrats, many of its leaders are co-opted by government to staff the new agencies or simply tire of battle, and support dwindles as the media turn to newer issues and the public assumes the problematic conditions are being taken care of by government. Efforts to revitalize the movement and avoid stagnation and co-option may lead to rancorous in-fighting and fragmentation, with "die-hard activists" disavowing those co-opted by government or seduced into working "within the system." These trends may result in the demise of the movement, as it disappears with little if any improvement in the problematic conditions that generated it. This "natural history" model seems to fit the rise-and-fall pattern of many social-problem movements (Mauss, 1975).[5]

A similar analysis of the problem-solving cycle has been offered by political scientists, who focus on the policy development and implementation stages. They argue that policies designed to solve social problems seldom succeed for two major reasons: First, interest groups (which successful social movements become at institutionalization) often achieve only symbolic victories, with government passing reassuring but essentially meaningless legislation. Second, even well-intended agencies are likely to fail, typically because they are captured by the very interests they were designed to regulate. This pattern of regulatory agency failure has been termed the "natural decay" model of government problem-solving efforts (see Sabatier and Mazmanian, 1980).

In short, social scientists posit a pattern in which social problems—such as "the environment"—are regularly discovered or created by activists, who are occasionally successful in getting the larger society to accept their definition of conditions as problematic and in need of amelioration. Such efforts are generally transitory and seldom fully successful, however, and generally experience a natural decline, often with little, if any, improvement in the problematic conditions. Has the environmental movement avoided this fate, as suggested by the success of the 1990 Earth Day anniversary, and, if so, why?

The Stages of Modern Environmentalism

The environmental movement was clearly institutionalized in the late 1960s and early 1970s, as signified by a flood of new groups at the national and especially the local levels, formalized media attention, and far-reaching legislation (Fessler, 1990). However, it appeared to lose steam fairly quickly. Within two or three years most organizations experienced slowed growth rates and several (especially at the local level) disappeared, public awareness and concern declined significantly, and explicitly anti-environmental counter-movements were being launched (Hays, 1987). It appeared that the movement had passed its peak and was experiencing a natural decline by the mid-1970s (see Albrecht, 1976). These trends accelerated during the pro-environmental Carter administration (1976–1980), as much of the public was lulled into thinking that environmental problems were being solved and many leading environmentalists were co-opted into the administration (Manes, 1990). Nevertheless, the strong organizational base did not fade away, and environmental protection continued to receive substantial if not consensual support from the public (Mitchell, 1989).

Ironically, the election of Ronald Reagan reversed these trends by stimulating a resurgence of environmental concern and activism. The anti-environmental orientation of his administration, highlighted by Department of Interior Secretary Watt and Environmental Protection Agency Director Gorsuch, provided environmental organizations with reason—and ammunition—for mobilizing opposition to his policies. Several of the national organizations were remarkably successful in recruiting new members, and their increased visibility and intensified lobbying helped stimulate some degree of Congressional opposition to administration policies.

In the process, the environmental movement not only avoided a demise but it experienced a major revitalization, gaining increased membership, new organizations (especially at the local levels), and renewed support from the public and policymakers.[6] This suggests that efforts by government to overtly repress or "de-institutionalize" a well-entrenched movement (as opposed to using more subtle "capture and co-optation" techniques) may backfire. By threatening environmentalists' hard-won "interest group" status, the Reagan administration rekindled the movement's zeal and activism.[7] However, although the movement successfully used the administration's hostility to its own benefit, it was largely unsuccessful in preventing Watt, Gorsuch, and their successors from halting nearly two decades of progress in federal environmental reform (Hays, 1987).

This clearly highlighted the weakness of relying on the governmental regulatory apparatus that the national organizations had worked so hard to create (Manes, 1990; Scarce, 1990).

The revitalization of environmentalism in the 1980s was sparked by opposition to the Reagan agenda and facilitated by the relatively high level of resources available to the environmental movement.[8] However, it was also stimulated by several underlying phenomena (Mitchell, 1989):

(1) The inherent and widespread appeal of environmental protection, stemming from the highly visible and increasingly threatening nature of environmental problems to virtually all segments of the population;
(2) the fact that progress in areas such as urban air quality was quickly offset by the emergence of new problems, often of a wider scale (e.g., acid rain) and more ominous nature (e.g., ozone depletion) than the older ones;
(3) the increasing societal recognition of continual, unanticipated environmental deterioration, resulting from the institutionalization of "environmental science" in government, academia, and especially environmental organizations themselves (and obviously stimulated by media attention); and
(4) the fact that environmental awareness had been institutionalized not only within the movement per se, but within many government agencies, scholarly associations, educational institutions, and even churches. Such trends have led Oates to argue that "a substantial degree of ecological consciousness has become a permanent part of the American value system" (Oates, 1989, p. 186; see also Milbrath, 1984; Paehlke, 1989).

Increased Diversity: Strength or Weakness?

Although environmentalism has clearly endured over the past two decades, with unintentional aid from its opposition, it nonetheless has changed substantially. The major change appears to be its vastly increased diversity. As Gottlieb noted, "By the end of the 1980s . . . environmentalism meant many different things to different groups and movements" (Gottlieb, 1990, p. 42). Although this diversity may lead to fragmentation, which Mauss (1975) sees as a precursor to the demise of a movement, we believe that it may prove to be an important strength of contemporary environmentalism. The contributions to this volume document and analyze major aspects of this diversity and, in the process, describe the evolution of environmentalism over the past two decades.

Chapter 2, by Mitchell, Mertig, and Dunlap, focuses on the largest environmental organizations, or the "national environmental lobby." It documents the vital role that the older conservation organizations played in generating environmentalism, the dramatic growth experienced by both the older and newer national organizations in the last 20 years, and the important role the national organizations took on in opposing the Reagan administration. It also discusses the difficulties created when these organizations professionalized to cope with large memberships and budgets and to maintain an effective presence in the nation's capital. Along with the insider's analysis provided later by McCloskey (a long-time President and now Chairman of the Sierra Club), Mitchell et al. document the problems that social-movement organizations encounter when they succeed in becoming large and entrenched interest groups.

Despite the homogenizing trends of professionalization and bureaucratization, the national environmental organizations have developed considerable diversity. Both Mitchell et al. and McCloskey note the purposeful specialization of these organizations in terms of their primary issues (air or water pollution, wildlife, energy conservation, etc.) and their basic strategies (lobbying, lawsuits, scientific research, and electoral politics); McCloskey also provides insight into their increasing collaboration with one another as well as with organizations outside the movement. Taken together, these two contributions provide a good understanding of how the major environmental organizations have evolved over the last 20 years.

A fundamental change in environmentalism since 1970 has been the rapid increase in the number and prominence of local grassroots organizations. These loosely coordinated volunteer groups typically develop in response to local environmental hazards that pose a threat to human health. Although epitomized by Love Canal, such groups have emerged in reaction to a wide range of problems besides toxic wastes, for example, existing or proposed landfills and waste incinerators. Freudenberg and Steinsapir provide a good overview of these local organizations, highlighting their differences from (and sometimes difficulties with) the national organizations. Besides being more likely to draw members from blue-collar and, increasingly, minority communities (Bullard, 1990)—thus helping environmentalism overcome its elitist image (Morrison and Dunlap, 1986)—the goals and tactics used by the local "not in my backyard" (NIMBY) campaigns typically differ from those used by the national organizations.

The recent emergence of grassroots environmentalism within minority communities is especially important, since racial and ethnic minorities traditionally have been wary of the environmental movement, fearing that it deflects attention from social justice concerns. Bullard and Wright describe how growing opposition to "environmental racism," the practice of locating "LULUs" (locally unwanted land uses) within minority communities, has led to the blending of environmental and equity concerns into an "environmental justice" movement. They review several cases of successful mobilization against LULUs within African-American communities, noting that pre-existing social action groups such as civil rights organizations are typically more influential than are the mainstream environmental organizations in stimulating such mobilization.

Another recent development is the emergence of a radical wing of environmentalism (although, admittedly, NIMBY campaigns sometimes use radical tactics). Fueled by a sense that the national organizations have failed in their efforts to reform the system (especially during the Reagan era), radical environmentalism calls for direct action (e.g., sit-ins) and illegal actions (e.g., "monkey-wrenching") if necessary (Manes, 1990; Scarce, 1990). Concerned primarily with wilderness and wildlife, many radical environmentalists draw inspiration from the philosophy of *deep ecology.* They see mainstream or reformist organizations (and most NIMBY groups) as shallow in their primary concern with human welfare, and they endorse a "biocentric" perspective that denies the favored status of *Homo sapiens.* Devall gives a good overview of both deep ecology and the organizational bases of radical environmentalism (Earth First!, Greenpeace, Sea Shepherd Society, and the Rainforest Action Network) and, in the process, delivers a denunciation of reformist environmentalism.

Although this volume focuses on the U.S. environmental movement, it is clear that environmentalism has been globalized within the past decade. Because environmental problems are increasingly transnational and often global in scale and therefore require international action, because such problems are in turn related to conditions within other nations (e.g., rainforest destruction in Brazil), and because U.S. environmentalists are often called on to assist environmental movements abroad (especially in the Third World), the major U.S. environmental organizations have adopted an increasingly global orientation (McCormick, 1989). In addition, international organizations such as Greenpeace have found considerable support in the United States. Caldwell offers a good overview of these developments, documenting how and why U.S. environmentalism has become globalized.

The foregoing analyses by academic-based writers are generally validated by McCloskey, a key participant in the development of U.S. environmentalism. As noted earlier, McCloskey generally concurs with Mitchell et al. on the evolution of national organizations such as the Sierra Club but offers additional insights—especially concerning the relations among the organizations—available only to an insider. His descriptions of ideology, strategy, and tactics reveal a good deal of diversity within mainstream environmentalism and also indicate that criticism from the radical and grassroots strands of environmentalism has provoked a good deal of soul-searching within the national organizations (see Scarce, 1990).

Supportive public opinion has proven a valuable resource for environmentalism, especially in the 1980s when it was mobilized to combat the Reagan administration; Dunlap analyzes trends in this support. Beginning by documenting the rapid increase of environmental awareness in the mid-1960s and leading up to the first Earth Day, Dunlap traces a quarter of a century of public concern with environmental problems. He documents the significant rebound in such concern in the 1980s and also describes the high level of public approval currently enjoyed by the environmental movement.

Overall, the contributions cover what we consider to be the most important dimensions of contemporary environmentalism, but admittedly not all of them. Probably most notable in their absence (because of space constraints) are eco-feminism, bioregionalism, animal liberation, and social ecology (the last one being a leftist brand of radical environmentalism that is highly critical of deep ecology and Earth First!; Bookchin, 1988). These strands or camps have been described and analyzed elsewhere and are also touched on by Devall (see Borrelli, 1987, 1988; Manes, 1990; and Scarce, 1990).

Even with these limitations, the contributions nonetheless reveal the diverse nature of contemporary environmentalism. Compared with 20 years ago, the movement has a much more diverse organizational base (from local to national and international), draws hard-core activists and volunteers from a wider range of social strata (especially more from the working class, minorities, and women), and addresses a far wider range of problems (from local health hazards to wildlife habitat to global ecosystem protection) with a greater range of tactics (from lawsuits to green consumerism to tree-spiking). This increased diversity has allowed environmentalism to fill (and create) many niches within our society and, as in nature, increased diversity may lead to greater resiliency in social movements.

Nevertheless, we cannot ignore the potentially fragmenting effect of such diversity. The critiques of mainstream environmentalism by Devall and by Freudenberg and Steinsapir, representing radical and grassroots environmentalism, respectively, reflect vast differences in the goals, tactics, and ideologies of these three major strands of the movement (to say nothing of the battles between deep ecologists, ecofeminists, and social ecologists). Indeed, many members of the various factions are eager to disassociate themselves from some of the other factions; for example, the mainstreamers who label radicals as "eco-terrorists," the deep ecologists who despise the reformers almost as much as polluters, and the NIMBY-ites who deplore those who value wildlife more than human lives. Given this diversity, one might argue that it is no longer appropriate to speak of *the* environmental movement.

Although cognizant of the significance of the important differences that have evolved among environmentalists over the past two decades, we must note that factions are not new (recall the Pinchot/Muir split within early conservationism). Furthermore, despite all of their differences, the various types of environmentalists share a recognition of the deteriorating state of the environment, a desire to halt such deterioration, and an opposition to those who foster it. What differentiates them are their diagnoses of the causes of the problems and their prescriptions for solving them. These are vast, indeed, and at times the resulting differences in strategies, tactics, and goals will no doubt be counterproductive. In the long run, however, we see this diversity as potentially enhancing the movement, providing it with more resources and personnel, and virtually guaranteeing that there will be a thirtieth Earth Day!

Conclusion: When Is a Movement Successful?

When environmentalism is judged against the typical social movement, we believe it is appropriate to argue that environmentalism has been a resounding success. By the first Earth Day in 1970 the environmental movement had achieved enormous visibility within our society and—albeit with some ups and downs—it has remained a viable sociopolitical force for more than two decades. However, social-problem movements develop for the purpose of solving problematic social conditions and, in this respect, environmentalism has not been atypical.

Many leading environmentalists, including McCloskey in this volume, have acknowledged that the movement has largely failed in its goal of protecting the quality of the environment. As Denis Hayes, key organizer for both the first and twentieth Earth Days, stated, "The world is in worse shape today than it was twenty years ago" (Hayes, 1990, p. 56). Of course, others are quick to point out that the situation would be far worse had the movement not been around. Although the primary purpose of this volume has been to examine environmentalism's success as a social movement, history will judge it in terms of its success in halting environmental deterioration rather than in simply avoiding its own demise.

Notes

1. The decision to limit the focus to the United States stemmed from space constraints, both in the original symposium and in this volume.

2. Historians differ in their views of the relative importance of the "utilitarian" and "preservationist" strands of conservationism at the time of the first wave as well as in subsequent waves of conservation (cf. Hays, 1959, and Nash, 1967; see also Paehlke, 1989, and Petulla, 1980).

3. Mitchell (1989) treats 1967, the year that the Environmental Defense Fund (the first "environmental-era" organization) was established, as the transformation point.

4. Some argued that environmental problems provided "the establishment" with a means of diverting the energies of youthful activism away from these more fundamental problems in U.S. society (see, e.g., Dunlap and Gale, 1972).

5. Mauss's model is compatible with the more recent "resource mobilization" perspective on social movements (Mauss, 1989), as illustrated by its parallels with Gale's (1986) effort to apply the latter to clarify the evolution of the environmental movement.

6. Mauss (1975) recognizes that under some conditions, revitalization may prevent movements from experiencing a demise, at least temporarily.

7. Gale's (1986) otherwise insightful analysis of the evolution of environmentalism, based on an examination of the interrelations between the movement, government agencies, and "anti-environmental" interests, neglects the fact that the environmental agencies were, in essence, captured by anti-environmental interests through Reagan's election. This suggests the need to

distinguish between environmental agencies, presidential administrations, and Congress in analyzing movement and "state" relationships (see, e.g., Sabatier and Mazmanian, 1980).

8. For example, in contrast with many movements, its core activists are well above average in education and are relatively affluent (Morrison and Dunlap, 1986), and foundations have been generous in support of its efforts (Mitchell, 1989). A more detailed comparison than we have been able to provide in this short essay of the resources of the environmental movement, relative to those of other movements, would provide valuable insights into environmentalism.

References

Albrecht, S. L. 1976. Legacy of the environmental movement. *Environment and Behavior,* 8:147–168.

Bookchin, M. 1988. Social ecology versus deep ecology. *Socialist Review,* 18 (July-Sept.):9–29.

Borrelli, P. 1987. Environmentalism at a crossroads. *Amicus Journal,* 9 (Summer):24–37.

———. 1988. The ecophilosophers. *Amicus Journal,* 10 (Spring):30–39.

Bullard, R. D. 1990. *Dumping in Dixie: Race, class, and environmental quality.* Boulder, CO: Westview.

Cahn, R., and P. Cahn. 1990. Did Earth Day change the world? *Environment,* September:16–20, 36–43.

Carson, R. 1962. *Silent spring.* New York: Houghton-Mifflin.

Dunlap, R. E., and R. P. Gale. 1972. Politics and ecology: A political profile of student eco-activists. *Youth and Society,* 3:379–397.

Fessler, P. 1990, January 20. A quarter-century of activism erected a bulwark of laws. *Congressional Quarterly Weekly Report,* 48:154–156.

Gale, R. P. (1986). Social movements and the state: The environmental movement, countermovement, and government agencies. *Sociological Perspectives,* 29:202–240.

Gottlieb, R. 1990. An odd assortment of allies: American environmentalism in the 1990s. *Gannett Center Journal,* 4:37–47.

Hart/Teeter. 1990. NBC News/*Wall Street Journal:* National Survey No. 6. Washington, DC: Author.

Hayes, D. 1990. Earth Day 1990: Threshold of the green decade. *Natural History,* April:55–58, 67–70.

Hays, S. P. 1959. *Conservation and the gospel of efficiency.* New York: Atheneum.

———. 1987. *Beauty, health, and permanence: Environmental politics in the United States, 1955–1985.* New York: Cambridge University Press.

Manes, C. 1990. *Green rage: Radical environmentalism and the unmaking of civilization.* Boston: Little, Brown.

Mauss, A. L. 1975. *Social problems as social movements.* Philadelphia: J. B. Lippincott.

———. 1989. Beyond the illusion of social problems theory. *Perspectives on Social Problems,* 1:19–39.

McCloskey, M. 1972. Wilderness movement at the crossroads, 1945–1970. *Pacific Historical Review,* 41:346–364.

McCormick, J. 1989. *Reclaiming paradise: The global environmental movement.* Bloomington: Indiana University Press.

Milbrath, L. W. 1984. *Environmentalists: Vanguard for a new society.* Albany: State University of New York Press.

Mitchell, R. C. 1989. From conservation to environmental movement: The development of the modern environmental lobbies. In *Government and Environmental Politics,* ed. M. J. Lacey, pp. 81–113. Washington, DC: Wilson Center Press.

Morrison, D. E., and R. E. Dunlap. 1986. Environmentalism and elitism: A conceptual and empirical analysis. *Environmental Management,* 10:581–589.

Nash, R. 1967. *Wilderness and the American mind.* New Haven, CT: Yale University Press.

Oates, D. 1989. *Earth rising: Ecological belief in an age of science.* Corvallis: Oregon State University Press.

O'Riordan, T. 1971. The third American conservation movement: New implications for public policy. *Journal of American Studies,* 5:155–171.

Paehlke, R. C. 1989. *Environmentalism and the future of progressive politics.* New Haven, CT: Yale University Press.

Petulla, J. M. 1980. *American environmentalism: Values, tactics, and priorities.* College Station: Texas A&M University Press.

Sabatier, P., and D. Mazmanian. 1980. The implementation of public policy: A framework of analysis. *Policy Studies Journal,* 8:538–560.

Scarce, R. 1990. *Eco-warriors: Understanding the radical environmental movement.* Chicago: Noble Press.

Twenty Years of Environmental Mobilization: Trends Among National Environmental Organizations

ROBERT CAMERON MITCHELL

Graduate School of Geography
Clark University
Worcester, MA 01610
USA

ANGELA G. MERTIG

Department of Sociology
Washington State University
Pullman, WA 99164-4020
USA

RILEY E. DUNLAP

Departments of Sociology and Rural Sociology
Washington State University
Pullman, WA 99164-4006
USA

Abstract *The evolution of national environmental organizations over the past two decades is analyzed, with special attention given to the major organizations that engage in lobbying. The rapid growth experienced by the older organizations and the emergence and subsequent growth of several newer organizations are described. An overview of the activities of these national organizations (including a description of their relationships with one another) is given. Next, the high level of public support for their goals, the unique characteristics of environmental issues, and the efficacy of direct mail recruitment techniques are highlighted as key causes of the organizations' growth and success. Finally, the organizational consequences of these trends, in the form of staff professionalization and bureaucratization, for the national organizations are examined in some detail.*

Keywords Direct mail recruitment, environmental lobby, environmental movement, environmental organizations, lobbying, public interest groups.

When most people think about the environmental movement they are likely to think first of large national environmental organizations such as the Sierra Club, the National Audubon Society, or the National Wildlife Federation. These organizations have played a critical role in the development and evolution of the environmental movement. They have several million dues-paying adherents; command multimillion-dollar budgets; employ corps of full-time lobbyists, lawyers, and scientists; and enjoy widespread support. Although such national organizations do not define the entirety of the environmental movement, they are clearly the most visible and often the most influential actors in environmental policy debates.

In this article we describe the evolution of the national environmental organizations, particularly how they have changed in the 20 years since the first Earth Day. We first describe twenty-five key national environmental organizations, distinguishing between those that formally engage in lobbying and those that emphasize other activities, and highlight their dramatic growth rates. We next describe their activities in some detail, emphasizing how the organizations have evolved as a result of their high degree of involvement in national policymaking. Because their success has depended on maintaining a high level of public support, we also examine how the national organizations have managed to mobilize such support, including the crucial importance that direct mail has played in allowing environmental organizations to recruit large memberships. Finally, we discuss some of the organizational consequences of relying on mass members who have only modest commitments to the organizations.

Evolution of the National Environmental Organizations

The modern environmental movement includes many different social-movement organizations—local, state, regional, national, and international—that seek to protect the environment. In addition, some national organizations, such as the Sierra Club and the National Audubon Society, have local and regional chapters with considerable influence. Although local, state, and regional organizations are often very influential in policy arenas, we restrict the discussion to the most prominent national-level environmental organizations.

Table 1 presents a list of twelve prominent national environmental organizations that engage in extensive lobbying activities and that can be considered the core of the national environmental lobby.[1] Not all national environmental organizations lobby within political arenas because such activities can cause organizations to lose their nonprofit status (Mitchell, 1989). Although these twelve organizations engage in other activities, such as education campaigns, research, and litigation, they are distinguished from other national environmental organizations by the fact that they openly lobby for the development and implementation of environmental legislation.

The seven organizations founded before 1960 constitute the conservation movement's important organizational contribution to the environmental movement. The three oldest organizations, the Sierra Club, the National Audubon Society, and the National Parks and Conservation Association, are products of the first conservation movement, which occurred during the Progressive Era. Today both the Sierra Club and the National Audubon Society have hundreds of local chapters, and both rank among the top three environmental organizations in size and influence.

Between the two world wars, three more national organizations were added to the conservation movement. Two of them were founded by sportsmen—the Izaak Walton League and the National Wildlife Federation. The third of the inter-war conservationist organizations, the Wilderness Society, arose out of concern for the preservation of the remaining wildlands outside the national parks system. A fourth organization, Defenders of Wildlife, came into being during the waning years of the conservation era. Although

Table 1

National Environmental Lobbying Organizations

Time Period/Organization	Year Founded	Membership (Thousands)							1990 Budget ($ Million)
		1960	1969	1972	1979	1983	1989	1990	
Progressive Era									
Sierra Club	1892	15	83	136	181	346	493	560	35.2
National Audubon Society	1905	32	120	232	300	498	497	600	35.0
National Parks & Conservation Association	1919	15	43	50	31	38	83	100	3.4
Between the Wars									
Izaak Walton League	1922	51	52	56	52	47	47	50	1.4
The Wilderness Society	1935	(10)	(44)	(51)	48	100	333	370	17.3
National Wildlife Federation	1936	—	(465)	(525)	(784)	(758)	(925)	975	87.2
Post–World War II									
Defenders of Wildlife	1947	—	(12)	(15)	(48)	63	68	80	4.6
Environmental Era[a]									
Environmental Defense Fund	1967	—	—	30	45	50	(130)	150	12.9
Friends of the Earth	1969	—	—	8	23	29	(30)	30	3.1
Natural Resources Defense Council	1970	—	—	6	42	45	105	168	16.0
Environmental Action	1970	—	—	8	22	20	(13)	20	1.2
Environmental Policy Institute	1972	Not a membership group.							
Total		123	819	1117	1576	1994	2724	3103	217.3

Note. Membership data are for individual members. In most cases the data come from the organization's membership person. Data in parentheses are estimates. Membership and budget data for 1990 are taken from Gifford (1990) for all organizations except National Parks and Conservation Association and Izaak Walton League. Data for last two are taken from Burek (1991).

[a]A recent addition to the national lobbying scene is Greenpeace Action. Founded in 1988, it is the lobbying arm of Greenpeace USA. Its membership overlaps considerably with that of Greenpeace USA (see Table 2).

initially its interest was in the welfare of individual animals, it gradually broadened to include wildlife habitat and endangered species.

Prior to the 1960s, the predominant concerns of these conservation organizations involved land and wildlife issues. A common feature of these first-generation issues (Mitchell, 1989) is that they involved threats to particular areas or species. An awareness of a new generation of environmental problems began to develop in the 1950s. These second-generation issues, in contrast with the earlier ones, are not necessarily site- or species-specific; they involve consequences that are often delayed or subtle; and their causes are typically difficult to prove. Rachel Carson's book, *Silent spring* (1962), called attention to the new type of problems by focusing on DDT and its systemic impact.

By the end of the 1960s, the conservation organizations had adopted both of the defining characteristics of the new environmental movement: the emerging ecological perspective and an agenda that included the second-generation issues (Mitchell, 1989). The Sierra Club and the other older organizations did not in any way abandon their commitment to the first-generation issues when they embraced environmentalism. The ecological perspective provided an intellectual framework for linking the first- and second-generation issues, enabling the older organizations to work on both kinds of issues and to cooperate with the newer environmental organizations, which tended to concentrate primarily on the second-generation issues.

When did the conservation movement become the environmental movement? A convenient though necessarily arbitrary demarcation for the transition from one to the other is 1967, the founding year of the Environmental Defense Fund (EDF), the first of the new breed of national environmental organizations. Both EDF, whose origins stem from the battle to ban DDT, and the National Resources Defense Council (NRDC) were assisted by Ford Foundation grants. In comparison with EDF, which at its founding was a membership organization dominated by scientists and engaged in litigation, NRDC was originally an environmental law firm run by lawyers. However, within a few years the two organizations became much more alike in their mix of scientists and lawyers (Mitchell, 1979b).

Friends of the Earth (FOE), founded by David Brower after he was fired as the Sierra Club's executive director for acting too independently of its elected Board of Directors, and Environmental Action, founded by the student organizers of Earth Day, were regarded as much more radical than the other major organizations because of FOE's uncompromising opposition to most forms of development, big business, and big energy and because of the "new left" brand of environmentalism espoused by Environmental Action.

The Environmental Policy Institute (EPI) was founded by a group of FOE's Washington lobbyists who tired of what they felt were inefficiencies of the FOE leadership and organization. These dissidents acquired some wealthy patrons whose support permitted them to avoid the distractions of having to serve a membership or publish a magazine and instead concentrate on lobbying.

As the nation entered the environmental decade of the 1970s, the twelve organizations in Table 1 dominated the movement's Washington presence. All except Environmental Action and the Defenders of Wildlife belong to an informal coalition of leaders called the "Group of 10" that meets periodically to discuss common strategies and problems (Stanfield, 1985, p. 1352). However, Environmental Action, along with the League of Conservation Voters and the Conservation Foundation,[2] tends to work closely with the coalition. A few years ago the coalition produced a volume entitled *An Environmental Agenda for the Future* (Cahn, 1985).

Growth of the National Environmental Lobby

Over the 20-year period from the first Earth Day in 1970 to the twentieth anniversary celebration in 1990, most of the national environmental lobbying organizations experienced tremendous growth. Their growth has been concentrated in four spurts, each attributable to a period of heightened public concern that environmental amenities were threatened. It is significant that the first growth spurt occurred in the years just before Earth Day, demonstrating that this event was as much the culmination of a process as it was the launching of a new movement. During this period, 1960 to 1969, the former conservation organizations' total membership increased almost sevenfold, from 123,000 to 819,000,[3] as they became more aggressive and outspoken about threats to scenic national treasures and wildlife.

The next membership surge was due to the publicity surrounding Earth Day, when environmental issues were firmly placed on the national agenda. In just three years, 1969 to 1972, nearly 300,000 new members were added to the national environmental ranks, a 38% overall increase. The older organizations, which helped mobilize Earth Day, were in the best position to reap the Earth Day harvest because they had the most name recognition, commanded the financial resources and expertise needed to conduct large direct mail campaigns, and were able to make appeals based on both first- and second-generation issues. As a result, the older organizations accounted for five out of every six new environmental organization members after Earth Day.

A number of observers believed environmentalism would soon fade after the 1970 Earth Day when the public realized how much a thorough environmental cleanup would cost. Although membership growth rates slowed after 1972, a majority of the organizations nonetheless experienced significant growth during the 1970s, especially the newer "environmental era" organizations. The organizations' overall growth clearly confounded those who labeled the movement a fad (Downs, 1972).

Any lingering doubts about the continuing viability of the environmental cause were erased when the Reagan administration's attacks on environmentalism stimulated a new influx of members in the 1980s (Mitchell, 1990). Although both first- and second-generation issues were affected by the Reagan administration's policies, the organizations most closely identified with the first-generation issues reaped the greatest harvest, thanks to the motivational appeal of wildlife and wilderness issues and the high visibility of James Watt's alleged misstewardship of the nation's resources as President Reagan's first Secretary of the Interior. The Wilderness Society grew by a phenomenal 144% between 1980 and 1983, the Sierra Club by 90%, and Defenders of Wildlife and Friends of the Earth by about 40% each. Subsequent growth has been strong, so that by 1989 these organizations counted a total membership of more than 2,700,000.

The most recent surge in membership occurred at the turn of the decade (1990), stimulated by the visibility of ecological problems ranging from toxic wastes, beach contamination, the Exxon Valdez oil spill, ozone destruction, and global warming, as well as by the substantial mobilization efforts these organizations made in conjunction with the twentieth Earth day celebration (Cahn and Cahn, 1990). Total memberships for these organizations grew substantially from 1989 to 1990, and exceeded 3 million by the twentieth Earth Day anniversary. Although there is some evidence that this surge has recently slowed or stopped because of the 1990–1991 recession (Lancaster, 1991), judging from past experience it would be surprising if the organizations suffered a significant decline in membership over the next decade.

Two factors seem to account for the national lobbying organizations' substantial growth in memberships. The first is the high level of public concern about their issue agenda, which created the potential for mass mobilization on behalf of environmental protection. Public opinion polls have recorded strong support, albeit with some ups and downs, for environmental protection from 1970 to 1990 (Dunlap, 1989; Mitchell, 1979a, 1980, 1990). Views that were once shared by a relatively small number of environmental activists before the first Earth Day have apparently diffused to the public at large. Of particular importance is the fact that environmental concern cuts across socioeconomic and political categories (Jones and Dunlap, 1989; Mitchell, 1979a). The result is that tens of millions of Americans with at least some college and with upper-middle-class incomes (the optimal target group for environmental direct mail lists) are sufficiently sympathetic to the organizations' goals that it is profitable to include them in "prospect lists" for direct mail campaigns.

Second, the direct mail form of mass mobilization has proven to be especially suited to recruiting environmental organization supporters. The types of issues touted by environmental organizations have particularly captivating characteristics: They have a broad scope of appeal, they are extremely visible, they continually recur, they enable individuals to feel a sense of empowerment when engaging in activity on their behalf, and people dread the loss of the amenities that environmentalists wish to protect. Direct mail, in addition to the constant media coverage of environmental problems, has combined to keep the public aware of and concerned about such problems (Mitchell, 1990).

The use of mail solicitations to recruit supporters for causes is a relatively new invention. The growing availability of computers in the 1960s greatly facilitated the use of direct mail, although many of the techniques in current use were not developed until the 1970s (Godwin, 1988). Direct mail was first used to recruit new members for environmental organizations in the late 1950s when Defenders of Wildlife and the National Parks and Conservation Association, among others, began to send out prospect mailings. By 1965 the National Audubon Society was mailing a million prospect letters a year to target lists acquired by exchange with other organizations or through list brokers, a rate of mailing it maintained until 1971 when it doubled the rate to two million pieces (Robert Bridges, National Audubon Society, interview on 23 September 1977).

The importance of this technology is that it provides virtually the only efficient way for people to learn about organizations' goals and to make contributions to them (Godwin, 1988). The major focus of the appeal letters is the need for action to combat a particular threat to the environment. The organization describes its activities to relieve the threat and asks for the prospective member's monetary contribution.[4] In the time period under review the appeal's credibility was enhanced by the abundance of publicity given to environmental issues by the mass media.

The use of direct mail recruitment lowered the effort required to contribute to environmentalism, in the sense that all a prospective member had to do was read the mail appeal, enclose a check in a prepaid envelope, and mail it in (Godwin and Mitchell, 1984). The amount requested in the mailings is, of course, a small amount for people with above-average incomes, the target group for environmental mailings. However, numerous small contributions can add up to impressive amounts of money, even after the costs of recruiting and serving the members are taken into account.

Not every organization was able to successfully exploit direct mail and grow. The Izaak Walton League is the only organization that did not seek a mass membership, keeping instead to its chapter structure and maintaining a constant membership size over the last two decades. Environmental Action and Friends of the Earth both suffered from chronic financial problems and an overly ambitious political agenda that made it difficult for them to invest heavily in direct mail recruitment. They also lacked the will and expertise to pursue a sophisticated direct mail program. Environmental Action was organized as a collective for most of this period, with shared decision making and equal salaries for all staff. FOE's charismatic leader, David Brower, was more skilled in developing program initiatives than in carrying them out within the organization's budget constraints. Budget problems, staff turnover, and ultimately a serious split in the organization weakened FOE in the 1980s when it might have caught the Reagan–Watt wave. By the late 1980s both Environmental Action and FOE were struggling to overcome their problems by means of new leaders, mergers with other like-minded organizations, and greater attention to management.

Nonlobbying National Organizations

Although they engage in a wide variety of pro-environmental activities, the organizations listed in Table 1 are distinguished by the large amount of time and resources they devote to lobbying for environmental policies. Other national organizations are notable for the fact that they do not explicitly lobby. Their activities include conducting research, litigation, education programs, grassroots organizing, land purchase and maintenance programs, and (in a few cases) even direct action on behalf of environmental goals. Such organizations, which vary considerably in their tactics and geographical and subject focus, include such disparate organizations as the World Wildlife Fund, National Toxics Campaign, League of Conservation Voters, and Greenpeace USA.

Table 2 presents a sampling of several of the most important of the many national (and international) nonlobbying organizations. These organizations represent some of the newer strains involved in the environmental movement. Although the organizations frequently overlap in the types of work they emphasize, some distinctions can be drawn between them.

Direct action has become a popular tactical alternative for some of the newer environmental groups. Although Environmental Action, as discussed later, has at times been supportive of such tactics, the first organization to pursue direct action as a primary strategy was Greenpeace. Its tactics, which have won Greenpeace an enormous following, have included plugging industrial effluent pipes and maneuvering "Zodiac" inflatable boats between whalers and whales. This style of direct action was taken a step further by Earth First! and the Sea Shepherd Society, both of which are not averse to damaging property (e.g., tree spiking, monkey-wrenching bulldozers, and sinking whaling ships) and using other forms of "ecotage." It is interesting that both groups were founded by disgruntled members of older, more established environmental organizations who felt that those organizations were not aggressive enough (see Devall in this issue for more detail).

The organizations in the land and wildlife preservation category pursue a concern that has been important to environmentalists since the conservation movement's foundation in the nineteenth century. Now, however, the focus reaches far beyond the United States. Of the five organizations listed, all except the Nature Conservancy focus primarily on other countries, especially on those with tropical rainforests (see Caldwell in this issue for more detail). In fact, Conservation International was founded by members of the Nature Conservancy who felt that the Conservancy's program was not sufficiently international (Gifford, 1990).

The growth of substantial grassroots coalitions to battle toxic waste sites and other local environmental hazards is an important new motif of modern environmentalism.

Table 2
Selected Nonlobbying Environmental Organizations

Type/Organization	Year Founded	1990 Membership (Thousands)	1990 Budget ($ Million)	General Program
Direct Action				
Greenpeace USA	1971	2300	50.2	Nonviolent direct action, campaigning, research on marine, toxics, and peace issues
Sea Shepherd Conservation Society	1977	15	0.5	Direct action for marine mammal protection
Earth First!	1980	(15)	0.2	Direct action for wilderness protection
Land & Wildlife Preservation				
Nature Conservancy	1951	600	156.1	Purchases land for preservation
World Wildlife Fund	1961	940	35.5	Research and conservation of international wildlife and tropical forests
Rainforest Action Network	1985	30	0.9	Promotes rainforest protection
Rainforest Alliance	1986	18	0.8	Promotes rainforest protection
Conservation International	1987	55	4.6	Promotes rainforest protection
Toxic Waste				
Citizen's Clearinghouse for Hazardous Waste	1981	7	0.7	Supports local anti-waste site groups
National Toxics Campaign	1984	100	1.5	Campaigns against toxic wastes and works with local groups
Other Major Organizations				
League of Conservation Voters	1970	55	1.4	Involved in electoral campaigns
Sierra Club Legal Defense Fund	1971	120	6.7	Conducts litigation
Earth Island Institute	1982	32	1.1	Various projects on environment and peace issues

Note. Information for this table is taken from Gifford (1990). Numbers in parentheses are estimates.

Table 2 lists two such national organizations, both of which promote grassroots activism on the issue of toxic waste. The Citizen's Clearinghouse was founded by Lois Gibbs, former Love Canal activist, and extends her organizing skills to local groups across the nation. The National Toxics Campaign is well versed politically and scientifically but somewhat more removed from the grassroots than is the Clearinghouse (see Freudenberg and Steinsapir in this issue).

National environmental organizations have become quite numerous in recent years, prohibiting an exhaustive listing of them, but the final category in Table 2 indicates some of the additional varieties of organizations.[5] Some organizations specialize in certain tactics. The Sierra Club Legal Defense Fund, an offshoot of the Sierra Club, engages exclusively in litigation, whereas the League of Conservation Voters specializes in electoral politics. The Earth Island Institute, founded by none other than David Brower after his split with Friends of the Earth over its move to Washington, DC, supports various projects (Gifford, 1990).

These nonlobbying organizations have also expanded in number and in size since the first Earth Day. Most notably, Greenpeace has mobilized a substantial following in the past few years. This is due to a tremendous surge in its direct mail efforts that enabled it to capitalize on a reputation for uncompromising and effective action and, some critics would say, to go down the path of corporatism and ossification (Gifford, 1990). Although we have not compiled membership growth data for the nonlobbying organizations, Tober's (1989) data for the World Wildlife Fund and the Nature Conservancy show growth trajectories similar to those of the lobbies.

Overall, it is clear that the national environmental organizations have prospered during the past two decades. Not only have earlier conservation-minded organizations grown in size since Earth Day 1970, but the number and diversity of national organizations have dramatically increased (even aside from the large number of regional, state, and local organizations that have emerged since 1970). Although it inevitably results in redundancy and competition, overall this diversity of organizations is perhaps the environmental movement's greatest political strength.

Activities of National Organizations

Three basic tactical options are available to the environmental organizations: education, direct action, and policy reform. Education was a primary function of most organizations from the 1930s to the 1950s. In the early 1960s, Rachel Carson's *Silent spring* (1962) and Barry Commoner's *Science and survival* (1963) played substantial roles in defining the second-generation issues and provided examples of how an educational approach using "informative science" could address these types of problems. Carson and Commoner each identified an environmental problem (pesticides and radioactive fallout, respectively) and provided a full briefing in lay language of the scientific issues involved and the changes required for a solution. Both placed the problem in an ecological perspective. Carson and Commoner and the environmentalists they inspired believed at first that informative science, by communicating the nature and seriousness of problems, could inspire spontaneous citizen pressure for reform. By the late 1960s most environmentalists realized that an *exclusively* educational approach to policy change was insufficiently aggressive, although educational campaigns continue to play an important role in the activities of most environmental organizations. Occasionally, as in the case of the Natural Resource Defense Council's recent campaign against the use of Alar on apples, such information campaigns have had a direct policy impact.

Direct action, conversely, has been regarded by the mainstream organizations as too aggressive, although Environmental Action was sympathetic with this approach in the early 1970s[6] and Greenpeace emphasized it before easing into the mainstream. Although Greenpeace, Sea Shepherd, and Earth First! practice their own versions of direct action, as discussed earlier, the mainstream organizations refrain from such tactics.

The preferred tactical choice of the national environmental organizations has been that of policy reform. To reform policy, the organizations engage in electoral campaigns, congressional and administrative lobbying, overseeing administrative decision making, and, when necessary, litigation. In all cases the national organizations are attempting to influence policy through traditional and legitimate channels, concentrating their efforts in Washington where Congress and federal agencies hold sway over environmental policies.

Recently many of the organizations entered electoral politics by establishing political action committees (PACs) to fund favored candidates, some of whom also received campaign aid from local environmentalist workers. Efforts of this sort were previously restricted to the nonpartisan League of Conservation Voters.

Lobbyists play a crucial role in pressuring Congress and the various government agencies involved with environmental issues to enact new laws and implement the existing ones. Particularly noteworthy is the large increase in the number of full-time environmental lobbyists in Washington. In 1969, before Earth Day, only two full-time lobbyists served the environmental movement. By 1975, the twelve organizations listed in Table 1 employed 40 lobbyists; a decade later the number of environmental lobbyists had swelled to 88 (Mitchell, 1989).[7] This growth reflects the fact that in the United States, "Lobbying is accepted as a normal, if not essential, arrangement for ensuring organized interests a major role in lawmaking" (Rosenbaum, 1991, p. 74).

Participation in administrative decision making refers to a wide range of activities by which environmentalists seek to influence the implementation of existing environmental laws. Laws typically leave a great deal of discretion to agencies in both setting standards and enforcing them. Agencies and affected parties subsequently engage in administrative proceedings to determine the implementation of the laws. During the Reagan era, the role that environmental organizations traditionally played in administrative decision making was almost completely circumvented (see, e.g., Vig and Kraft, 1984).

Prior to the 1960s, environmentalists and citizens' groups were not granted standing to participate in administrative proceedings of this kind because the government agencies themselves were seen as sufficiently representative of the public interest. However, administrative law in the United States has now moved toward a system of *interest representation* (in no small part because of court victories by environmentalists) whereby *all* interested parties are granted standing (Mitchell, 1989; Stewart, 1975). This has given the environmental lobby the opportunity to press for strong standards and strict enforcement. They have done this by critiquing the environmental impact statements that are mandatory for major government projects, commenting on proposed regulations, petitioning for regulatory action when agencies have been too dilatory, participating in scientific advisory committees, providing information to the agencies, and testifying at administrative hearings.

After its inception as an environmentalist tactic in the mid-1960s, litigation quickly became one of the most important weapons in the environmentalist arsenal. Litigation is intimately related to administrative decision making because the courts are the environmentalists' last resort when they are unable to get an administrative agency to fulfill what the environmentalists consider to be its legal duty. In the United States the courts are uniquely powerful actors in the public policy process (Rosenbaum, 1991).

With the exception of litigation and explicit electoral support for candidates, none of these tactics is new, having been used by the conservation movement at one time or another. What is new is the scale of the environmental movement's advocacy, its sophistication, and especially its continuity. The national organizations now have the capacity to follow key issues from initial proposals to legislative enactment to the typically protracted, yet crucial implementation stage, in which an agency develops and enforces regulations designed to make the law work.

This general policy reform strategy is compatible with the environmental organizations' mass-membership form of mobilization, in which members delegate authority to the organizations and provide them with the resources necessary for lobbying and litigating actively. Changes in the rules of standing, as discussed earlier, as well as changes in the tax laws and regulations (which have made it possible for organizations to engage in noneducational advocacy without fear of losing their nonprofit tax status) were both pivotal in allowing environmental organizations to engage in political advocacy (Mitchell, 1989). Another significant factor was the professionalization of the organizations and their staff, permitting a much higher degree of continuity and staff expertise on issues than the organizations had previously possessed.

Institutionalization of Advocacy: Professionals and Specialists

Up to the early 1960s, U.S. conservation organizations such as the Sierra Club, the Wilderness Society, and the National Audubon Society were relatively small organizations of citizens whose forays into environmental policy making were limited to occasional lobbying. Much of the administrative workload was carried by volunteers. The number of paid staff concerned with policy issues was small, and many organizations were dominated by leaders who possessed charismatic qualities and exerted strong personal leadership (Langton, 1984; McCloskey, 1972). When advocacy campaigns were undertaken, the organizations often relied on volunteers for legal advice, lobbying, and congressional testimony. Fund-raising, personnel, accounting procedures, and both substantive and financial planning were often ad hoc. Taken together, the national organizations had only a few paid lobbyists, no paid legal staff for litigation, and no full-time scientists for research.

Organizations of amateurs such as these were fully capable of dealing with first-generation issues (Fox, 1981), but the second-generation issues were far more numerous, diverse, and complex, and their resolution often took years of sustained effort. Dealing with them required a high level of legal and scientific expertise. It was no accident, therefore, that the earlier amateurism was rapidly transformed into professionalism by the early 1970s, a professionalism whose hallmark was the sizable cadre of lobbyists, lawyers, and scientists employed full-time by the national environmental organizations. Volunteer experts continue to be used, but now their efforts are coordinated for the most part by the staff experts.

The pressures for professionalization converged from three directions. First, as the pace and complexity of legislation increased, the need to track its evolution and implementation through the bureaucratic labyrinth made a full-time professional advocacy staff more crucial. This need was greatest in Washington, so that the organizations could work effectively with the federal government. Organizations headquartered in other parts of the country either established Washington offices or moved their headquarters to the capital. Second, the organizations' rapid increase in membership stretched their management capacities. When membership growth slowed after the early 1970s and expected revenues failed to materialize, the ensuing financial crunch exposed the weakness of some of the organizations' financial planning and control. Since then, the need for professional management and expertise has become more and more urgent. Third, to keep their nonprofit tax status and satisfy the reporting requirements of nonprofit regulators, the organizations were forced to adopt complex organizational arrangements and maintain detailed records of their expenditures by purpose (see, e.g., Davies, Irwin, and Rhodes, 1984; Langton, 1984, for discussions of perceived need for improved leadership and management within environmental organizations).

The result was a shift from an essentially amateur management to a professional form of advocacy characterized by paid staff, planning exercises, budgets, and financial control. The dramatic increase in the number of Washington environmental lobbyists has already been noted. In addition, legal and scientific specialists were hired in increasing numbers. The organizations formalized their financial operations and hired professional fund-raisers. In 1984–1985, more than half of the organizations hired new executive heads (Stanfield, 1985), and many boards placed management ability near the top of the list of the characteristics they sought (Thompson, 1985). By the end of the 1970s, many of the lobbyists who had been hired a decade before were still at their posts or—another sign of professionalism—had taken better-paying jobs with the Carter administration or with other environmental organizations. Their accumulated experience was an important resource for the organizations.

The transition from amateur to professional lobbies was aided by the continued societal consensus about the need for environmental protection. This consensus reduced the movement's need to rely on charismatic leaders to engender public support and minimized the organizational inefficiency and instability that often accompanies this type of leadership (Langton, 1984). Their boards of directors (some elected by the memberships in competitive elections, most not) have played a crucial role in ensuring organizational continuity when leadership problems occurred. Over the past decade, for example, the boards of the Wilderness Society, Environmental Defense Fund, and Defenders of Wildlife, among others, have fired executive directors.

The environmental organizations in the United States, particularly the environmental law organizations (EDF and NRDC), have succeeded in institutionalizing a "counterscience" capability. They have access to sufficient scientific and technological expertise and information so that they can evaluate the scientific basis for environmental policies and, if necessary, dispute the scientific and technical issues within the appropriate policy arenas. The use of science by environmentalists in cases such as these may be termed *advocacy science* (Mitchell, 1979b).

Those practicing advocacy science marshall scientific evidence in support of a particular policy conclusion, such as banning the use of fluorocarbons in aerosol sprays, adopting a particular numerical limit for the number of porpoises killed by U.S. tuna fisherpeople, or banning Alar in apple production, and then actively press those claims in relevant policy forums. The fundamental aim of the advocacy is to ensure that technology is not permitted to inflict serious harm on the environment or human health. Because those who benefit materially from a given technology may be counted on to promote and defend it in administrative and judicial settings, environmentalists tend to find themselves in adversary relationships with industry scientists as each seeks to discredit the scientific evidence and arguments used by the other (Mitchell, 1979b).

Because they have the requisite in-house legal and scientific expertise, the environmental law organizations are most capable of participating in the administrative and litigative arenas. As mentioned earlier, EDF and NRDC are the two major organizations of this type, but the Sierra Club has its allied law organization, the Sierra Club Legal Defense Fund, and the National Wildlife Federation has an in-house legal staff. EDF and NRDC are staffed by lawyers and scientists, with a ratio of about two to three lawyers to each scientist with a Ph.D. Each has some 15 to 20 full-time staff scientists, whose specialties range from biochemistry to geology.

The national environmental organizations frequently practice the politics of ad hoc coalitions, in which several organizations (including nonenvironmental ones, such as labor unions, whenever possible) unite in a formal or informal coalition to work together on a given issue. Because the issues are complex and numerous, each organization tends to develop areas of specialization, and an informal division of labor results (e.g., EDF concentrates on toxic chemicals, wetlands, water quality, and power generation, whereas NRDC focuses on air pollution, nuclear power, and solid waste). Coordination of efforts is further increased by the fact that most of the organizations are now headquartered in Washington (see McCloskey in this issue for more detail).

What Success Has Meant for the National Organizations

A mail-recruited membership is not necessarily a highly committed one (Godwin and Mitchell, 1984). Furthermore, large national memberships mitigate against effective member input into organizational decision making. With the possible exception of the Sierra Club, members of the national environmental organizations do not exert direct influence on organizational decision making. Most often the organizations' staffs are responsible only to self-perpetuating boards of directors. Nevertheless, though most members of these organizations lack an effective direct voice in decision making, they possess an indirect voice through their voluntary memberships and contributions, which they can stop at any time.

The commitment to recruit a mass membership was not lightly made by the older organizations, whose boards of directors were often wary of the financial risks and potential organizational consequences. Nor was the implementation of the technology of direct mail uniformly smooth. Besides requiring and promoting staff professionalism, in those organizations with chapters, mass recruitment tended to strengthen the autonomy and influence of the headquarters staff relative to chapter officers. These trends are clearly illustrated in the experiences of the National Wildlife Federation and the Sierra Club, especially in the increased strength of the national offices relative to the National Wildlife Federation's state affiliates and the Sierra Club's local chapters as the national memberships expanded dramatically (Allen, 1987; Cohen, 1988).

There has been a clear trend toward increased professionalization among the leaders and staff of the national environmental organizations over the past two decades. The necessity of managing large budgets and complicated, multi-faceted programs made possible by the recruitment of large numbers of new members has resulted in managerial skills becoming more highly valued than are the visionary goals of charismatic leaders (Langton, 1984; Stanfield, 1985). Increased professionalization also carries with it the dangers of routinization in advocacy, careerism on the part of staff members, and passivity on the part of volunteers, all of which have been detected in the national organizations (Langton, 1984; also see McCloskey in this issue). An inevitable result has been that the national organizations are increasingly criticized as being bureaucratic and unresponsive to members, too eager to engage in political compromise, and more concerned about their budgets than about the state of the environment. It is not surprising that many of the new environmental organizations founded during the 1980s are a response to a perceived lack of aggressive advocacy by the older organizations. Thus, the very resources that have made the environmental lobby a significant force in national politics have also generated strains within the movement.

Conclusion

Compared with most U.S. social-movement organizations, the national environmental organizations have been enormously successful over the past two decades. They have not only survived but have experienced remarkable membership growth. The older ones completed the transition from conservationism to environmentalism and directly or indirectly contributed to the emergence of numerous new organizations. Taken together, these organizations constitute a potent force in national policy-making, and their stature reflects the fact that environmentalism has evolved from a loosely coordinated social movement to a cohesive public interest lobby (Berry, 1977). The increased political clout of the national environmental lobby is, however, accompanied by the inherently conservatizing pressure to play by the "rules of the game" in the compromise world of Washington, DC. We can therefore expect continued criticism of "the nationals" from radical environmentalists (see Devall in this issue) and continued tensions between the national organizations and local grassroots groups (see Freudenberg and Steinsapir in this issue).

Given the high level of public support for environmental quality and the probable continued visibility of threats to that quality, it is likely that despite their problems the national environmental organizations will continue to attract members and resources for the foreseeable future. Their success in balancing the organizational demands posed by large memberships and budgets against the necessity of achieving real gains in environmental protection will have a major impact on both the environmental movement and the physical environment in future years.

Notes

1. For a more detailed discussion of the organizations in Table 1, see Mitchell (1989).

2. See Table 2 for information on the League of Conservation Voters. The Conservation Foundation was founded in 1948 and does not recruit members but often provides support for other organizations (Stanfield, 1985). For example, the Foundation conducted a study of the need for training within environmental organizations (Davies, Irwin, and Rhodes, 1984).

3. Because some individuals belong to more than one organization, "total memberships" in Table 1 exceed the total number of people who belong to at least one organization.

4. The direct mail method of recruiting mass members (Godwin, 1988; Godwin and Mitchell, 1984) involves bulk mailing (at cheap rates) of appeal letters to lists of prospects. Because many people contribute to more than one organization, the best prospect lists for a particular organization are the membership lists for organizations that share its interests. Once recruited, members receive mailings asking for special contributions, as well as annual reports that seek to persuade them to renew their memberships for another year. The cost of annual memberships ranges from $15 to $40, and many members donate additional money. Although renewal rates are much higher than the 0.5% to 2% response rates to new prospect mailings, achieving sufficient recruitment and

retention to expand or even to maintain a given membership size is inherently problematic because it is dependent on repeated individual decisions to contribute.

5. For a more complete list of national and international environmental organizations, interested readers should consult the latest edition of the National Wildlife Federation's *Conservation Directory,* updated annually. It should be emphasized that Tables 1 and 2 include the vast majority of the most important national/international environmental organizations (see Gifford, 1990).

6. It viewed with favor the efforts of an elusive activist, known as "the Fox," who for a time embarrassed corporations by gestures such as depositing noxious wastes from their discharge pipes in corporation headquarters. Environmental Action even sponsored a book entitled *Ecotage!* (Love and Obst, 1972), although they have not endorsed the recent wave of direct action.

7. For a good overview of environmental lobbyists and the types of activities in which they engage, see Berry's (1977) study of public interest lobbyists. All of the organizations listed in Table 1, except the National Parks and Conservation Association, were included in his study.

References

Allen, T. B. 1987. *Guardian of the wild: The story of the National Wildlife Federation, 1936–1986.* Bloomington: Indiana University Press.

Berry, J. 1977. *Lobbying for the people: The political behavior of public interest groups.* Princeton, NJ: Princeton University Press.

Burek, D. M. 1991. *Encyclopedia of associations: National organizations of the U.S.: Vol. 1.* Detroit, MI: Gale Research, Inc.

Cahn, R., ed. 1985. *An environmental agenda for the future.* Washington, DC: Island Press.

Cahn, R., and P. Cahn. 1990. Did Earth Day change the world? *Environment,* 32:16–20, 36–43.

Carson, R. 1962. *Silent spring.* Greenwich, CT: Fawcett Crest.

Cohen, M. P. 1988. *The history of the Sierra Club: 1892–1970.* San Francisco: Sierra Club Books.

Commoner, B. 1963. *Science and survival.* New York: Viking Press.

Davies, J. C., F. H. Irwin, and B. K. Rhodes. 1984. *Training for environmental groups.* Washington, DC: Conservation Foundation.

Downs, A. 1972. Up and down with ecology—the "issue-attention cycle." *The Public Interest,* 28:38–50.

Dunlap, R. E. 1989. Public opinion and environmental policy. In *Environmental politics and policy: Theories and evidence,* ed. J. P. Lester, pp. 87–134. Durham, NC: Duke University Press.

Fox, S. 1981. *John Muir and his legacy: The American conservation movement.* Boston: Little, Brown.

Gifford, B., and eds. 1990. Inside the environmental groups. *Outside,* 15:69–84.

Godwin, R. K. 1988. *One billion dollars of influence: The direct marketing of politics.* Chatham, NJ: Chatham House Publishers.

Godwin, R. K., and R. C. Mitchell. 1984. The impact of direct mail on political organizations. *Social Science Quarterly,* 65:829–839.

Jones, R. E., and R. E. Dunlap. 1989, August. Socio-demographic correlates of public support for environmental spending: Trends from 1973 to 1987. Revision of paper presented at the annual meeting of the Rural Sociological Society, Seattle, WA.

Lancaster, J. 1991, February 15. War and recession taking toll on national environmental organizations. *Washington Post,* p. A-12.

Langton, S. 1984. *Environmental leadership: A sourcebook for staff and volunteer leaders of environmental organizations.* Lexington, MA: D.C. Heath.

Love, S., and D. Obst. 1972. *Ecotage!* New York: Pocket Books.

McCloskey, M. 1972. Wilderness movement at the crossroads, 1945-1970. *Pacific Historical Review,* 41:346–361.

Mitchell, R. 1979a. *Silent spring*/Solid majorities. *Public Opinion* 2:16–20, 55.

——— . 1979b. *Since* Silent spring: *The institutionalization of counter-expertise by the United States environmental law groups.* (Resources for the Future Discussion Paper D-56). Washington, DC: Resources for the Future, Inc.

——— . 1980. *Public opinion on environmental issues: Results of a national opinion survey.* Washington, DC: President's Council on Environmental Quality.

——— . 1989. From conservation to environmental movement: The development of the modern environmental lobbies. In *Government and environmental politics: Essays on historical developments since World War Two,* ed. M. J. Lacey, pp. 81–113. Washington, DC: Wilson Center Press.

——— . 1990. Public opinion and the Green Lobby: Poised for the 1990s? In *Environmental policy in the 1990s,* ed. N. J. Vig and M. E. Kraft, pp. 81–99. Washington, DC: Congressional Quarterly.

Rosenbaum, W. 1991. *Environmental politics and policy,* 2d ed. Washington, DC: Congressional Quarterly.

Stanfield, R. L. 1985, June 8. Environmental lobby's changing of the guard is part of movement's evolution. *National Journal,* pp. 1350–1353.

Stewart, R. 1975. The reformation of American administrative law. *Harvard Law Review,* 88:1669–1813.

Thompson, G. 1985. New faces, new opportunities: The environmental movement goes to business school. *Environment,* 27(4):6–30.

Tober, J. 1989. *Wildlife and the public interest: Nonprofit organizations and federal wildlife policy.* New York: Praeger.

Vig, N. J., and M. E. Kraft, eds. 1984. *Environmental policy in the 1980s: Reagan's new agenda.* Washington, DC: Congressional Quarterly.

Not in Our Backyards: The Grassroots Environmental Movement

NICHOLAS FREUDENBERG
CAROL STEINSAPIR

Community Environmental Health Center
School of Health Sciences
Hunter College
City University of New York
New York, NY 10010
USA

Abstract *In the last decade, environmental activism in local communities has grown dramatically. Unlike the national environmental organizations, which are predominantly white and middle class, local environmental groups draw their members from a broad cross-section of class and occupational categories. Members of minority groups are active as well, and women often have leadership roles. Unlike the nationals, grassroots groups emphasize protection of public health rather than the environment and often mistrust government and scientists. Local groups have forced cleanup of contaminated dump sites, blocked proposed facilities, and developed support for a preventive approach to environmental contamination. Functioning as self-help groups, they assist individuals and communities to cope socially and psychologically with toxics disasters. Critics who deride the "not in my back yard" (NIMBY) perspective of some local groups ignore their contributions to public health and the fact that many groups develop a "not in anyone's back yard" (NIABY) perspective.*

Keywords Environmental health, environmental movement, environmental racism, grassroots environmental organizing, NIABY, NIMBY.

In 1978, residents of the Love Canal neighborhood in Niagara Falls, New York, propelled the issue of toxic chemical pollution onto the front pages of national newspapers, onto the agendas of Congress and state legislatures, and into the minds of the U.S. people. Since that time, thousands of other communities have organized to force the cleanup of a contaminated site, to shut down a polluting factory, or to prevent the construction of a facility perceived to be hazardous. In this article we trace the origins of the grassroots environmental movement, describe its characteristics, compare it with the more traditional environmental organizations, and assess the movement's accomplishments.

Although the Love Canal crisis captured the first national headlines, the seeds of the grassroots environmental movement were planted in an earlier period. Several factors help to explain the dramatic expansion of local environmental activity in the late 1970s and throughout the 1980s.

First, the growth of the petrochemical industry after World War II helped to ensure that factories, waste sites, and transportation routes were distributed throughout the country. Between 1940 and 1980, production of synthetic organic chemicals increased from less than 10 billion pounds a year to more than 350 billion (Davis and Magee, 1979). Each year, 1500 new compounds are added to the estimated 70,000 chemicals already in use by 1980 (Conservation Foundation, 1987). In 1988, the U.S. General Accounting Office estimated that there were between 130,000 and 425,000 contaminated waste sites in this country, although less than 1500 sites are currently included on the Superfund National Priorities List (Montague, 1989). Data reported under the new federal community right-to-know law indicate that U.S. industry releases more than 2.7 billion pounds of toxic chemicals into the atmosphere each year, including more than 360 million pounds of suspected carcinogens (Environmental Protection Agency [EPA], 1989; "These polluters," 1989). According to the EPA, more than 90% of U.S. citizens have measurable quantities of chemicals such as styrene, ethyl phenol, ethyl benzene, or toluene in their bodies (Montague, 1989). It is hardly an exaggeration to say that nearly every community has some potential environmental health hazard.

Second, public awareness of environmental hazards increased, at least among some sectors of the population. Rachel Carson's *Silent spring* (1962), Barry Commoner's *The closing circle* (1971), the educational work associated with Earth Day in 1970, the activities of the antinuclear power movement and the earlier campaigns against aboveground nuclear testing led by scientists and physicians, and the emergence of national environmental organizations such as the Natural Resources Defense Council and the Environmental Defense Fund all contributed to a growing awareness of the relationship between pollution of the environment and human health. Although these efforts primarily reached educated U.S. citizens, they helped create a different public attitude toward the environment. By 1980, a national public opinion poll showed that 90% of the respondents would not want a nuclear power plant near their home, and 40% said they would not want to live within 5 miles of a toxic waste dump (Jackson and Wright, 1981). Ten years earlier, almost no one would have worried about these hazards.

Third, the social movements of the 1960s and 1970s had helped create widespread awareness of new strategies and tactics for social change. The civil rights and antiwar movements had shown that community organizing, demonstrations, civil disobedience, and political education could lead to changes in government policy and programs. They had also created a cadre of experienced activists who later played important advisory or leadership roles in some grassroots environmental struggles. In the 1970s, the movement against nuclear power and citizen action groups such as Fair Share and ACORN helped develop leadership, build organizational structures, and create links with legal and scientific experts that were used directly by grassroots environmental groups. Although it is always difficult to trace the precise impact of one movement on another, it seems clear that the organizational forms and strategies used by the grassroots environmental movement borrowed heavily from the earlier social movements.

Description of the Grassroots Environmental Movement

By 1990, grassroots environmentalism had evolved into a loosely structured movement with three overlapping but distinct levels of organization: community-based groups, regional or statewide coalitions, and national organizations. Local community organizations constitute the foundation of the movement. Precise numbers are not available;

however, the National Toxics Campaign, a Boston-based organization that provides technical assistance to grassroots groups around the country, now has a mailing list of 1300 groups, more than double their list of only two years earlier ("Grass-roots," 1989). The Citizen's Clearinghouse for Hazardous Waste, another national group, reports that it works with 7000 grassroots environmental groups struggling to protect their communities against some perceived hazard (Gibbs, 1989b).

What are the characteristics of these groups? Surveys and case histories by Brown (1990), Edelstein (1988), Edelstein and Wandersman (1987), Freudenberg (1984a, 1984b), Levine (1982), and others permit some preliminary generalizations. Many are formed by a small group of people who are directly affected by a perceived health hazard in their community. In one survey, for example, nearly half of the responding organizations reported that health concerns led them to organize the group (Freudenberg, 1984a). In some cases, victims or families of victims play an important role in the organization (e.g., Woburn, Love Canal, Rutherford; Brown, 1990; Freudenberg, 1984b; Gibbs, 1982; Levine, 1982). This can give the group a moral legitimacy that is a powerful organizing tool.

Typically, groups begin by attempting to document a hazard and link it to a current or potential health problem, such as a cluster of cancer cases or a series of adverse reproductive outcomes. This activity often leads to extensive interactions with scientists, public health officials, and sometimes lawyers (Freudenberg, 1984a, 1984b). Problems addressed by local groups include toxic dumps, pesticide spraying, air pollution, contaminated water supplies, radioactive wastes, nuclear plants, and proposals to build garbage incinerators and hazardous waste disposal facilities (Freudenberg, 1984a).

Once the group members decide that there is an association between an actual or potential exposure and human health, they usually attempt to convince government (or, less commonly, the industry itself) to clean up, shut down, or abandon plans to build a new facility. The initial objective is usually to correct a specific problem, not to effect broader policy changes. Failure to achieve goals at this level can lead groups to enter the legal system by filing lawsuits or to move into the political arena where they lobby for legislation, endorse candidates, or propose ballot initiatives (Freudenberg, 1984b).

The cumulative effect of these activities can lead to dramatic changes in political consciousness for activists. Their faith in government and business is often reduced, and their willingness to move beyond mainstream activities such as research, lobbying, and electoral activity can lead to more dramatic tactics such as picketing the homes of key opposition figures or holding sit-ins that block trucks bringing equipment to construct new facilities (Freudenberg, 1984b). At Love Canal, Lois Gibbs and other residents held two EPA officials "hostage" for several hours; two days later, President Jimmy Carter declared Love Canal a disaster area, making the residents eligible for relocation assistance (Gibbs, 1982). When people feel that the health of their families is at stake they are sometimes willing to take actions that are illegal or that violate community standards of acceptable behavior. Often, the environmental struggle becomes a dominant passion in the lives of participants.

Members of these local groups include a broad cross-section of class and occupational categories. Women are heavily represented in both membership and leadership. Many groups have experienced community or political activists among their founding members, but it is a distinguishing characteristic of this movement that new leaders, often housewives and mothers with no previous organizing experience, also emerge (Freudenberg, 1984a; Freudenberg and Zaltzberg, 1984; Levine, 1982). Almost all local groups depend entirely on volunteers to carry out their work.

Unlike the more traditional environmental movement represented by organizations such as the Sierra Club and the National Audubon Society, grassroots environmentalism has also attracted African-Americans, Native Americans, and Latinos to its ranks. In Warren County, North Carolina, a primarily African-American group struggled to block dumping of PCB-contaminated soil in a landfill; on the Pine Ridge Reservation, Women of All Red Nations sought to force cleanup of contaminated water and land; in California, Mothers of East Los Angeles organized a Mexican-American community to block construction of an oil pipeline through its neighborhood (Freudenberg, 1984b; "Mothers group," 1989).

This involvement by minority groups reflects the fact that minorities are disproportionately exposed to environmental health hazards (Bullard and Wright, 1986–1987). A study by the Commission for Racial Justice of the United Church of Christ demonstrated that commercial hazardous waste disposal facilities are more likely to be located in communities with substantial minority populations (Commission for Racial Justice, 1987). Blacks have higher blood levels of carbon monoxide and pesticides, and black children have a rate of lead poisoning six times that of white children (Kutz, Yobs, and Strassman, 1977; National Center for Statistics, 1984; Radford and Drizd, 1982). There is growing recognition in minority communities that these patterns reflect broader inequities of economic and political power in our society, and some activists have begun to use the term "environmental racism" to link environmental issues to the struggle for social justice and racial equality. As a result, environmental organizers in minority communities often raise questions about political power as well as about public health (Freudenberg, 1984b).

A number of factors spur local groups to join together in networks or coalitions: a desire for allies, a need not only for basic scientific and technical information but also for an analysis that explains the underlying causes of the complex problems they face, and a growing recognition of the interrelationships between scientific and political issues. These larger groupings have diverse structures and strategies and seem to have been studied less than the grassroots organizations. Examples include the Grass Roots Environmental Organization (GREO) in New Jersey, the Citizens Environmental Coalition (formerly the Toxics in Your Community Coalition) in New York State, the Silicon Valley Toxics Coalition in California, and Texans United. Some networks, such as GREO and the Citizens Environmental Coalition, link groups working on diverse issues within a defined geographic region. Others, such as Work on Waste, a New York State coalition formed to fight mass-burn incineration and promote recycling, and the New York Coalition for Alternatives to Pesticides, link groups working on the same issue in different communities. They serve to educate members about relevant scientific and political issues, provide a forum for exchanging experiences and developing new strategies, and enable local groups to advocate jointly for new policies and programs at the state level.

Finally, a handful of national organizations have emerged since 1981 both to provide scientific, legal, and political assistance to the smaller groups and to lobby and advocate at the national level. These include the previously mentioned Citizen's Clearinghouse for Hazardous Wastes, founded in 1981 by Lois Gibbs, a leader of the struggle at Love Canal; and the National Toxics Campaign, with its headquarters in Boston. These organizations have paid staff and scientific and legal consultants. They organize national conferences for activists, offer leadership training, publish newsletters and manuals, provide technical assistance to local groups, and develop policy papers and legislative initiatives. They also channel grassroots concerns to legislators and lobby for legislation that protects the environment and strengthens the role of neighborhood organizations. In some cases, regional organizers work directly with local groups to help them in their campaigns. Similar services are provided on a regional basis by groups such as Citizens for a Better Environment (in the Midwest and in California) and the Southwest Research and Information Center.

Shared Perspectives Among Grassroots Environmentalists

Although the grassroots environmental movement is hardly a homogeneous grouping, its organizations, activists, and leaders generally share certain principles and beliefs. We will later discuss how these beliefs distinguish this movement from the older, more established conservation and environmental organizations.

First, grassroots environmentalists strongly believe in the right of citizens to participate in making environmental decisions. Advocacy for community right-to-know legislation, support for citizen enforcement provisions in federal and state legislation, and the insistence by many local groups on playing a role in deciding how to clean up a contaminated site or whether to close down a facility illustrate this concern. This emphasis on the process as well as the content of environmental decision making reflects the traditional democratic aspirations of at least some community organizing efforts in the United States (Boyte, 1980).

The emphasis on citizen participation in decision making also appears to reflect a mistrust of government based on direct experiences with public officials and agencies. In one survey of grassroots environmental groups, for example, 45% of the respondents claimed that government agencies had blocked their access to needed information (Freudenberg, 1984a, 1984b). If government cannot be trusted to safeguard health, then community activists want to ensure that they can participate in making decisions and can represent their interests directly. For those environmentalists with experience in earlier social movements, the focus on grassroots empowerment links the struggles to clean up the environment with a larger effort to correct social injustices and redistribute political power.

Second, human health, rather than environmental esthetics, wilderness preservation, or other such issues, is the primary concern of most grassroots environmentalists. The motivating factor in most of their struggles appears to be a desire to protect their own health and the health of their families and future generations against some perceived threat. Critics of the movement sometimes charge that concerns about health are used as a smokescreen to cover fears that construction of facilities such as incinerators and hazardous waste dumps will damage property values. Although this is undoubtedly true in some cases, there are many other situations in which people persist in fighting an environmental hazard despite corporate threats that jobs will be lost or despite opposition by neighbors who want to keep evidence of pollution quiet in order to protect property values (Brown, 1990; Freudenberg, 1984b; Levine, 1982). In any case, critics who accuse grassroots environmentalists of using health issues as a cover for other motives implicitly accept the legitimacy of health concerns as a motivation and moral justification for action.

A third characteristic of grassroots environmental groups is their ambivalent attitude toward scientific and technical expertise. On the one hand, several investigators reported close and positive relationships between activists and scientists (Brown, 1990; Freudenberg, 1984b; Levine, 1982). Nearly every group has had some interaction with technical experts, and many report scientists to be their most important source of information. On the other hand, both surveys and anecdotal reports suggest widespread mistrust of scientists and public health officials. What explains this apparent discrepancy? Activists seem to distinguish between scientists employed by industry or government, whose jobs depend on refuting the claims of damage to human health, and those who are perceived to be willing to use their expertise to support the social goals of the movement or a specific community group. As Montague (1989) observed, some scientists have become more like lawyers, culling facts from a body of evidence to make a case for their client. Implicitly, if not explicitly, environmental activists reject the image of science as a neutral force that pursues the truth no matter what its consequences.

This skepticism about science also reflects the fact that government and industry often seek to convert what are fundamentally political issues into scientific and technical questions that require expert study rather than democratic decisionmaking. As a result, environmental activists may see government and industry scientists as fronts for the real decision makers. Grassroots activists also resist attempts to define environmental problems as purely technical ones because they would prefer to confront their adversaries on political terrain, where their skills and strengths can be used to maximum advantage.

Fourth, implicit in much of the movement's actions is a challenge to the prevalent belief that economic growth is good per se and ultimately benefits everyone. Community groups opposing a new facility that they believe will threaten their health are unwilling to trade their own welfare for alleged benefits to society as a whole. Thus, grassroots groups have implicitly contested the assumptions of cost–benefit analysis by asking who pays the costs and who gets the benefits. Without necessarily recognizing the anticapitalist implications of their position, these groups have also questioned the rights of corporations to make decisions without community input if those decisions will have health and social consequences.

Comparison with the National Environmental Organizations

These characteristics of the grassroots environmental movement distinguish it from larger environmental organizations such as the National Wildlife Federation, the Audubon Society, and the Sierra Club, as well as from newer groups such as the Environmental Defense Fund (EDF) and the Natural Resources Defense Council (NRDC). Although there are important differences among these national organizations, they also tend to share some attributes. The older organizations were traditionally concerned with the protection of land and wildlife. In the 1960s and 1970s, they began to add toxics to their agendas, but retained their commitment to conservation issues. Groups founded in this period, such as NRDC and EDF, from the outset devoted considerable resources to air pollution, water pollution, and other toxic chemical problems, but human health frequently received less emphasis than general environmental quality. From the perspectives of many grassroots activists, the national organizations still seem to be more interested in protecting threatened animal species from extinction than in protecting children from toxic pollutants in their own backyards.

The primary constituency of the national organizations is white, middle-class Americans; their leaders and staff are almost exclusively white ("Environmental groups," 1990; Morrison and Dunlap, 1986). Few of these organizations have reached out to or attracted working-class Americans, African-Americans, or Latinos. The national groups tend to focus most of their work on national legislation or litigation and see concrete changes in law or policy as the most important outcomes of their work. Many have local and state chapters, but the resources of the organization are concentrated in their headquarters. Their staff have a large proportion of scientists and lawyers who spend a considerable amount of time in court, Congress, or scientific meetings. These experts interact regularly with industry and government experts, with whom they share a professional training and an understanding of the "rules of the game." Unlike grassroots activists, these professionals have a stake in preserving their credibility with other experts and with government decision makers. This may encourage a willingness to compromise on particular issues to preserve valuable relationships for the long term (Montague, 1989). As a result, grassroots groups sometimes perceive the nationals as remote, overly legalistic, and too willing to accommodate to industry's concerns (Borrelli, 1988; Montague, 1989).

Although organizations such as NRDC and the EDF may share the grassroots skepticism about the benefits of uncontrolled economic growth, their perspective appears to be somewhat different. They are heavily influenced by the concepts of ecological limits and sustainable growth and emphasize environmental and economic arguments for sustainable agriculture and energy conservation rather than emphasizing public health rationales. Although the differences in perspective reflect different priorities, the goals are not mutually exclusive and there may be room for the development of common agendas on these issues.

In the last decade, national environmental organizations have devoted greater attention to health issues like lead poisoning and air pollution, and several of them have provided technical assistance to community organizations involved in local environmental struggles (Borrelli, 1988). Despite these changes, however, the significant differences outlined above continue to distinguish these two segments of the environmental movement.

Accomplishments of the Grassroots Environmental Movement

Because the national environmental organizations and the grassroots environmental movement have overlapping agendas and strategies, it is not possible to make conclusive distinctions between their separate accomplishments. In this section, however, we attempt to describe some of the achievements for which the grassroots environmental movement is primarily responsible.

First, the movement has successfully forced the cleanup of contaminated dump sites, blocked the construction of garbage incinerators and hazardous waste disposal facilities, won bans on aerial spraying of pesticides, and forced corporations to upgrade pollution control equipment. According to Lois Gibbs, no new hazardous waste disposal facilities have been built since 1978 in communities where there was organized opposition to such facilities (Gibbs, 1989a).

It seems clear that community organizations have contributed to specific improvements in public health as a result of such activities. Although it is seldom possible to quantify the impact of their interventions, it is reasonable to conclude that some cases of cancer, birth defects, or other health problems have been prevented. That we lack the scientific tools to evaluate the contribution of grassroots organizations to the health of the public should not lead us to ignore its significance.

Second, community organizations have forced corporations to consider more closely the environmental consequences of their actions. In 1983, for example, Dow Chemical decided to withdraw the herbicide 2,4,5-T from the market. The negative publicity from community struggles in the Pacific Northwest, Arkansas, and elsewhere, and the threat of product liability suits made continued production unprofitable (Freudenberg, 1984b). In a similar vein, some polluting corporations targeted by citizen action groups decided that it would be less costly politically and economically to install new pollution control devices than to engage in ongoing battles with community residents (Freudenberg, 1984b).

Third, the cumulative impact of separate local struggles has created new political and economic pressures for a preventive approach to environmental contamination. By blocking construction of hazardous waste facilities and new garbage incinerators—and thereby closing off new disposal options—grassroots groups have collectively forced industry and government to look for ways to reduce waste generation at the source. This will require new production technologies, substitution of safer products for more dangerous ones, and increased recycling.

This preventive approach to pollution control has begun to be embodied in legislation as a result of grassroots pressure. Examples include Proposition 65 in California, which bans any discharges of certain toxic chemicals into drinking water, and the recent legislation in Massachusetts mandating that corporations reduce their use of toxic chemicals. What is significant about these legislative acts is their emphasis on prevention of pollution and their clear focus on human health. The new focus on pollution prevention contrasts sharply with the traditional regulatory approach, which entails setting legally allowable limits on pollution that are achieved by end-of-the-pipe pollution controls.

Fourth, the grassroots environmental movement has won legislative victories that expand the rights of citizens to participate in environmental decision making. One example is the 1986 Superfund community right-to-know provisions, which enable citizens to learn the names and quantities of hazardous chemicals that local factories store or emit into the water and air. Another example is the Technical Assistance Grant program, also part of the 1986 Superfund reauthorization, which offers up to $50,000 to local groups at Superfund sites to hire their own technical advisers (Gibbs, 1989a).

Fifth, grassroots environmental organizations have had an important impact on communities affected by toxic disasters. Recent research suggests that the psychological consequences of toxic chemical disasters may be as important as the damage to physical health (Bachrach and Zautra, 1985; Brown, 1990; Edelstein, 1988; Levine, 1982). Similar to a self-help group, the community organization provides social support and mutual help, helps victims understand and channel their grief and anger, and offers a vehicle for making the experience of the disaster meaningful (Brown, 1990; Edelstein, 1988). The process of protest can help develop a sense of neighborhood, which can in itself be therapeutic and can help the community combat future threats to its welfare (Unger and Wandersman, 1985).

For the activists who participate in organizing a community campaign, the process can teach new skills, build self-confidence, and create new and meaningful friendship networks (Edelstein, 1988; Freudenberg, 1984b; Freudenberg and Zaltzberg, 1984; Levine, 1982). Some individuals, such as Lois Gibbs of Love Canal, have gone on to become national leaders in the environmental movement. It should be noted, however, that activists also report disruption of family life, divorce, and isolation from previous friends and neighbors as possible outcomes of their struggles (Freudenberg, 1984b; Freudenberg and Zaltzberg, 1984; Gibbs, 1982; Levine, 1982).

Sixth, as mentioned earlier, the grassroots movement has brought environmental concerns and action to working-class and minority Americans. By raising health concerns and by linking environmental issues to struggles for social justice and equality, the grassroots environmental movement has created the potential for a cross-class movement with a broader agenda, more diverse constituencies, and a more radical critique of contemporary society than that of the national environmental organizations. Whether this fledgling movement can realize this potential remains to be seen.

Finally, the grassroots environmental movement has influenced how the U.S. people think about the environment and public health. A *New York Times* national poll in June 1989 showed that 80% of the population agreed with the statement, "Protecting the environment is so important that requirements and standards cannot be too high, and continuing environmental improvements must be made regardless of the cost;" in 1981, early in the Reagan administration, only 45% of respondents had agreed with this position ("Grass-roots groups," 1989, p. A-1.). Media coverage of local environmental struggles as well as citizens' direct participation in these campaigns apparently have played an important role in increasing public support for environmental protection.

The Future: From NIMBY to NIABY

Members of the grassroots environmental movement are often attacked for their NIMBY stand, which is characterized as a narrow, self-interested negativism that ignores obligations to the larger society. This critique fails to acknowledge the substantial public health benefits of NIMBYism as described earlier. In the absence of rational planning for economic development that minimizes environmental damage, the NIMBY phenomenon is often a constructive contribution to the general welfare.

Critics of NIMBY groups also fail to acknowledge that many local groups move beyond simple nay-saying to support socially constructive alternatives that express a NIABY philosophy (Collette, 1989). Thus, many groups opposing the construction of garbage incinerators become advocates for recycling and waste reduction measures such as packaging controls.

This transcendence of NIMBY by NIABY is a critical step if the grassroots movement is to make a substantial contribution to resolving the pollution problems confronting the United States and the world at large. The Citizen's Clearinghouse for Hazardous Wastes and the National Toxics Campaign have made an important contribution by working to foster this broader perspective while strongly supporting local NIMBY struggles.

As the grassroots movement begins to adopt the "NIABY" philosophy, it will increasingly be drawn into the national policy arena. This will raise the possibility of greater collaboration with the traditional national environmental organizations. Such a collaboration could strengthen both parties. The nationals can provide technical and financial resources and expertise on Capitol Hill. The grassroots movement, with its focus on specific toxics problems that directly threaten the health of community residents, can mobilize masses of people who may not define themselves as environmentalists. Whether these complementary assets and overlapping agendas can overcome the ideological and tactical differences between these two sectors of environmentalism remains an open question. The answer will strongly influence the shape of the environmental movement in the coming decade.

References

Bachrach, K., and A. Zautra. 1985. Coping with a community stressor: The threat of a hazardous waste facility. *Journal of Health and Social Behavior,* 26:127–141.

Borrelli, P. 1988. Environmentalism at a crossroads. In *Crossroads: Environmental priorities for the future,* ed. P. Borrelli, pp. 3–25. Washington, DC: Island Press.

Boyte, H. 1980. *The backyard revolution: Understanding the new citizen movement.* Philadelphia: Temple University Press.

Brown, P. 1990. *No safe place: Toxic waste crisis and childhood leukemia in Woburn, Massachusetts.* Berkeley: University of California Press.

Bullard, R. D., and B. H. Wright. 1986–1987. Blacks and the environment. *Humboldt Journal of Social Relations,* 14:165–184.

Carson, R. (1962). *Silent spring.* Boston: Houghton Mifflin.

Collette, W. 1989. Organizing toolbox: Taking a stand. *Everyone's Backyard,* 7(Winter):4–5.

Commission for Racial Justice. 1987. *Toxic waste and race in the United States: A national report on the racial and socio-economic characteristics of communities with hazardous waste sites.* New York: Author.

Commoner, B. 1971. *The closing circle.* New York: Knopf.

Conservation Foundation. 1987. *State of the environment: A view toward the nineties.* Washington, DC: Author.

Davis, D. L., and B. H. Magee. 1979. Cancer and industrial chemical production. *Science,* 206:1356–1358.

Edelstein, M. 1988. *Contaminated communities: The social and psychological impacts of residential toxic exposures.* Boulder, CO: Westview.

Edelstein, M., and A. Wandersman. 1987. Community dynamics in coping with toxic contaminants. In *Neighborhood and community environments,* ed. I. Altman and A. Wandersman, pp. 69–112. New York: Plenum Press.

Environmental groups told they are racist in hiring. 1990, February 1. *New York Times,* p. A-20.

Environmental Protection Agency. 1989. *The toxics-release inventory: A national perspective, 1987.* Washington, DC: U.S. Environmental Protection Agency, Office of Toxic Substances.

Freudenberg, N. 1984a. Citizen action for environmental health: Report on a survey of community organizations. *American Journal of Public Health,* 74:444–448.

———. 1984b. *Not in our backyards! Community action for health and the environment.* New York: Monthly Review Press.

Freudenberg, N., and E. Zaltzberg. 1984. From grassroots activism to political power: Women organizing against environmental hazards. In *Double exposure: Women's health hazards on the job and at home,* ed. W. Chavkin, pp. 246–272. New York: Monthly Review Press.

Gibbs, L. 1982. *Love Canal: My story.* Albany: State University of New York Press.

———. 1989a. The Movement on the move. *Everyone's Backyard,* 7(Summer):1, 3.

———. 1989b. Together we can win justice. *Grassroots Convention 89 songbook.* Arlington, VA: Citizen's Clearinghouse for Hazardous Waste.

Grass-roots groups show power battling pollution close to home. 1989, July 2. *New York Times,* p. A-1.

Jackson, A., and A. Wright. 1981. Nature's banner. *Progressive,* October, p. 26.

Kutz, F. W., A. R. Yobs, and S. C. Strassman. 1977. Racial stratification of organochlorine insecticide residues in human adipose tissues. *Journal of Occupational Medicine,* 19:6219–6229.

Levine, A. G. 1982. *Love Canal: Science, politics, and people.* Lexington, MA: Lexington Books.

Montague, P. 1989. What we must do: A grass-roots offensive against toxics in the '90s. *Workbook,* 14:90–113.

Morrison, D. E., and R. E. Dunlap. 1986. Environmentalism and elitism: A conceptual and empirical analysis. *Environmental Management,* 10:581–589.

Mothers group fights back in Los Angeles. 1989, December 5. *New York Times,* p. A-32.

National Center for Health Statistics, 1984. Blood lead levels for persons 6 months to 74 years of age: United States, 1976–1980. *Vital Health Statistics,* No. 79. Hyattsville, MD: Author.

Radford, E., and T. Drizd. 1982. Blood carbon monoxide levels in persons 3–74 years of age:

United States, 1976–1980. *Advance Data,* No. 76, p. 8. (Hyattsville, MD: National Center for Health Statistics.)

These polluters give America a real pounding. 1989, June 20. *Daily News,* (New York), p. 10.

Unger, D., and A. Wandersman. 1985. The importance of neighbors: The social, cognitive, and affective components of neighboring. *American Journal of Community Psychology,* 13:139–169.

The Quest for Environmental Equity: Mobilizing the African-American Community for Social Change

ROBERT D. BULLARD

Department of Sociology
University of California
Riverside, CA 92521

BEVERLY H. WRIGHT

Department of Sociology
Wake Forest University
Winston-Salem, NC 27106

Abstract *There is considerable evidence that pollution and other forms of environmental degradation take a heavy toll on African-American communities. African Americans, however, have not been attracted to mainstream environmental groups. Grassroots environmental groups—with their emphasis on environmental justice—have attracted a larger following within African-American communities. The 1980s have seen the struggle for environmental equity take on a civil rights meaning. Using preexisting social justice and grassroots organizations for leadership, black community residents have begun to challenge government and private industries that would turn their areas into the nation's dumping grounds for all types of health-threatening toxins.*

Keywords African Americans, environmental justice, protests, racism, toxics, waste

Much research has been devoted to analyzing environmental movements in the United States (Buttel and Flinn, 1978; Dunlap, 1987; Gale, 1983; Hays, 1987; Mitchell, 1979; Morrison, 1980, 1986; Morrison and Dunlap, 1986; Schnaiberg, 1980, 1983; Van Liere and Dunlap, 1980). Despite this wide coverage, there is a dearth of material on the convergence of environmentalism and social justice advocacy. Nearly two decades ago, Gale (1972) compared the civil rights movement and the environment movement. The modern environmental movement has its roots in the civil rights and antiwar movements of the late sixties (Humphrey and Buttel, 1982). Student-activists who broke away from the civil rights and antiwar movements formed the core of the environmental movement in the early 1970s.

There is a substantial body of literature on grassroots environmental groups

This paper was presented at the annual meeting of the American Sociological Association, San Francisco, California, 9–13 August 1989.

(Freudenberg, 1984; Freudenberg and Steinsapir, 1990; Gottlieb and Ingram, 1988). However, little research has been conducted on African-American, Latino, and Native-American grassroots environmental groups such as the Gulf Coast Tenants Organization (New Orleans), Mothers of East Los Angeles, Concerned Citizens of South Central Los Angeles, Southwest Organizing Project (Albuquerque), Toxic Avengers of Brooklyn, West Harlem Environmental Action, or Native Americans for a Clean Environment (Oklahoma). A special issue of *Environmental Action* highlighted the fact that the time is long overdue for the nation to move "beyond white environmentalism" (Truax, 1990).

This paper analyzes environmental activism within five African-American communities in the South. The issues that are addressed include: (1) factors that shape environmental mobilization; (2) the level of convergence between environmental justice and social equity goals; (3) the source of environmental leadership; and (4) types of dispute resolution strategies used.

The Environmental Justice Movement

The civil rights movement has its roots in the southern United States. Southern racism deprived blacks of "political rights, economic opportunity, social justice, and human dignity" (Bloom, 1987, p. 18). However, racism is by no means limited to any one region of the country. The environmental justice movement for African Americans is centered in the South, a region where marked ecological disparities exist between black and white communities. Many of these disparities were institutionalized by laws and public policies during the "Jim Crow" era. African-American communities systematically became the "dumping grounds" (Bullard, 1990, p. 43) for all kinds of locally unwanted land uses (LULUs).

The literature is replete with studies documenting the disproportionate environmental burden borne by African Americans [Bullard, 1990; Bullard and Wright, 1986, 1987a; Commission for Racial Justice (United Church of Christ), 1987; Gianessi, Peskin, and Wolff, 1979; Jordon, 1980; Kazis and Grossman, 1983; Kruvant, 1975; McCaull, 1975; U.S. General Accounting Office, 1983]. These ecological disparities were highlighted in the 1983 Urban Environment Conference whose theme was "taking back our health." The conference was held in New Orleans and was one of the first national forums that brought together "Third World" people in this country and progressive whites to talk about environmental justice and coalition building. This broad-based group of civil rights activists, organized labor leaders, and grassroots environmental activists formed a loose alliance (Urban Environment Conference, Inc., 1985, p. 29). Cooperative action between social justice and environmental groups was seen as one of the best strategies to weaken the hold of "job blackmail"—the threat of job loss or plant closure—on working class and communities of color (Kazis and Grossman, 1983).

A growing number of African-American grassroots environmental groups and their leaders have begun to adopt confrontational strategies (e.g., protests, neighborhood demonstrations, picketing, political pressure, and litigation) similar to those used in earlier civil rights disputes. These activists advocate a brand of environmentalism that attempts to address disparate impact and equity issues. Documentation of civil rights violations has strengthened the move to make environmental quality a basic right of all individuals (Bullard and Wright, 1987a, 1987b).

Social justice advocate Reverend Ben Chavis defines many of the ecological inequities within the African-American community as direct results of "environmental racism.'' The privileges of whites (access to a clean environment) are created and institutionalized at the expense of people of color (Commission for Racial Justice, 1987, p. *x*). Thus the practice of targeting urban ghettos or rural blackbelt communities for noxious facilities (i.e., hazardous waste landfills, incinerators, paper mills, garbage dumps, and other polluting industries) is seen as another expression of institutional racism.

The U.S. General Accounting Office (1983) observed a strong relationship between the locations of off-site hazardous waste landfills and the racial and socioeconomic status of surrounding communities. The GAO looked at eight southern states that comprised the federal Environmental Protection Agency's (EPA's) Region IV (Alabama, Florida, Georgia, Kentucky, Mississippi, North Carolina, South Carolina, and Tennessee). The government study identified four off-site hazardous waste landfills in the region. African Americans made up a majority of the population in three of the four communities where the off-site hazardous waste landfills are located. The fourth site is located in a community where 38% of the population is African American. In 1990, only two off-site hazardous waste landfills were operating in Region IV, and both of these sites are located in African-American communities. African Americans, who make up only 20% of the region's population, continue to shoulder a heavier toxic waste burden than any other group in the region.

The Commission for Racial Justice's (1987) ground-breaking report *Toxic wastes and race* clearly shows that African-American communities and other communities of color bear a heavier burden than society at large in the disposal of the nation's hazardous waste. Race was the most potent variable in predicting the location of uncontrolled (abandoned) and commercial toxic waste sites in the United States.

The nation's total hazardous waste disposal capacity was 127,989 acre-feet in 1987 (Commission for Racial Justice, 1987). In 1987, three of the nation's largest commercial hazardous waste sites were located in African-American or Latino communities. These three sites—Chemical Waste Management site in Emelle, Alabama (black); Rollins Environmental Services in Scotlandville, Louisiana (black); and Chemical Waste Management site in Kettleman City, California (Latino)—accounted for 51,070 acre-feet of the disposal capacity, or 40% of the nation's total hazardous waste disposal capacity in 1987 (Commission for Racial Justice, 1987).

The first national protest by African Americans against environmental racism came in 1982 after the mostly black Warren County, North Carolina, was selected as the burial site for 32,000 cubic yards of soil contaminated with highly toxic polychlorinated biphenyls, or PCBs. The soil was illegally dumped along the roadways in 14 North Carolina counties in 1978. African-American civil rights activists, political leaders, and local residents marched in protest demonstrations against the construction of the PCB landfill in their community. More than 400 demonstrators were jailed. Although the protests were unsuccessful in halting the landfill construction, they marked the first time that blacks mobilized a nationally broad-based group to protest environmental inequities and the first time that demonstrators had been sent to jail for protesting against a hazardous waste landfill (Bullard and Wright, 1986; Geiser and Waneck, 1983, pp. 13–17).

The Attraction of Environmental Justice

There is no single agenda or integrated political philosophy binding the hundreds of environmental organizations found in the nation. The types of issues that environmental organizations tackle can greatly influence the types of constituents they attract (Gale,

Table 1
Type of Environmental Groups and Issue Characteristics
Likely to Attract Black Community Residents

	Type of Environmental Group			
Issue Characteristics	Mainstream	Grassroots	Social Action	Emergent Coalition
Advocate safeguards against environment-economic tradeoffs	–	–/+	+	+
Focus on inequality and civil rights	–	–/+	+	+
Endorse the politics of direct action	–/+	+	+	–/+
Seek political empowerment of "underdog" groups	–	–/+	+	–/+

Source. Adapted from Gale (1983).
Key: –, group is unlikely to have characteristic; +, group is likely to have characteristic; –/+, group may have characteristic in some cases.

1983, pp. 182–186). The issues that are most likely to attract the interests of African-American communities are those that have been couched in some civil rights or equity framework (Table 1). They include those that (1) advocate safeguards against environmental blackmail with a strong pro-jobs stance, (2) focus on inequality and civil rights, (3) endorse the politics of direct action, and (4) seek the political empowerment of "underdog" groups.

Mainstream environmental organizations such as the Sierra Club, National Wildlife Federation, Natural Resources Defense Council, and the Environmental Defense Fund have had a great deal of influence in shaping the nation's environmental policy (Gale, 1983). These organizations, however, have not had much success in attracting poor and working-class persons or the large urban underclass (which is burdened with both pollution and poverty) in the nation's central cities or the rural southern blackbelt (Morrison and Dunlap, 1986; Taylor, 1989). Many of these individuals do not see the mainstream environmental movement as a vehicle that is championing the causes of political "underdog" groups (Bullard and Wright, 1987b; Jordon 1980; Taylor, 1982).

Many white grassroots environmental groups, especially those that deal with the toxics issue, have begun to broaden their base of issues to include social justice. These groups may be community-based, regional, or national in scope (Freudenberg and Steinsapir, 1990). The Citizen's Clearinghouse for Hazardous Waste (CCHW), founded by Lois Gibbs, leader of the Love Canal fight, and the National Toxics Campaign are probably the best known national grassroots environmental organizations that have embraced the goal of environmental justice. The motto of CCHW is "People united for environmental justice." Grassroots environmental groups have had more success than their mainstream counterparts in attracting people of color.

Within the African-American community, environmental groups often emerge from established social action organizations. For example, leadership within the African-

American community has deep roots in the church and social justice organizations. These indigenous institutions have led the opposition against school and housing segregation, job discrimination, and a host of other institutional barriers. Social change within many of these communities occurred as a direct result of the actions taken by indigenous leaders of civic clubs, neighborhood associations, community improvement groups, and an array of antipoverty and antidiscrimination organizations.

It is not an accident that the Reverends Ben Chavis of United Church of Christ's Commission for Racial Justice, Joseph Lowery of the Southern Christian Leadership Conference (SCLC), and Jesse Jackson of the National Rainbow Coalition have taken leadership roles on the toxic waste and race issue. The issue is couched in a manner that retains much of the civil rights movement flavor. For example, Jesse Jackson's "toxics tour" of African-American communities in April of 1990—part of the Earth Day 1990 celebration—is a classic case in which civil rights and environmental justice agendas converged. The tour began with a series of speeches held at Atlanta University Center, which is a cluster of five historically black institutions that have a long history of producing African-American leaders (e.g., Martin Luther King, Jr.).

In some instances the struggle for environmental reform in African-American communities has come from alliances and coalitions between environmentalists (both mainstream and grassroots), social action advocates, and organized labor (Pollack and Grozuczak, 1984). These fragile alliances operate from the position that social justice and environmental quality are compatible goals. Although these groups are beginning to formulate agendas for action, mistrust still persists among them as a limiting factor (Bullard and Wright, 1987b). The coalitions are often biracial with membership cutting across class and geographic boundaries. Compositional factors may engender less group solidarity and sense of control among African-American members, compared to the indigenous social action or community-based grassroots environmental groups in which African Americans are in the majority and therefore make the decisions.

Case Studies from the Southern United States

The African-American community is bombarded with all kinds of environmental stressors. Our goal was to select varied noxious facilities—municipal landfill, hazardous waste landfill and incinerator, and lead smelter facilities—and then gauge public opposition to these facilities. On the surface, municipal landfills may seem nonthreatening because they receive household garbage. However, sanitary landfills routinely accept household waste—some of which is highly toxic—and many receive illegally dumped hazardous waste. The threat (whether real or perceived) of living next to a toxic waste dump and incinerator has been well publicized. Similarly, the ill-effects of lead poisoning have been known since the Roman era.

The heightened militancy among African Americans on environmental issues served as a backdrop for studying a special brand of environmentalism. This analysis centered on five African-American communities that were involved in environmental disputes during 1979–1987: Houston's Northwood Manor neighborhood (Texas), the West Dallas neighborhood (Texas), Institute (West Virginia), Alsen (Louisiana), and Emelle (Alabama). Although the communities have different histories, they share the common action of challenging the notion that social justice and environmental concern are incompatible goals.

The analysis is based on in-depth interviews conducted with a total of fifteen opinion leaders who were identified through the "reputational" approach and who were active in the local environmental disputes. The interviews were conducted in the summer of 1988 and were supplemented with archival documents and newspaper articles, editorials, and feature stories on the disputes. The data addressed the following: (1) issue crystallization (e.g., each opinion leader was asked how he/she defined the local dispute); (2) citizen opposition tactics used; (3) type of leadership that spearheaded the citizen opposition; (4) methods used to resolve the dispute; and (5) outcome of the dispute.

Houston, Texas

Houston in the 1970s was dubbed the "golden buckle" of the Sunbelt. In 1982 it became the nation's fourth largest city, with 1.7 million inhabitants. Houston is the only major U.S. city that does not have zoning, a policy that has contributed to haphazard and irrational land-use planning and infrastructure chaos (Babcock, 1982; Bullard, 1984, 1987; Feagin, 1985, 1987). Discriminatory facility-siting decisions allowed the city's black neighborhoods to become the "dumping grounds" for Houston's municipal garbage (Bullard, 1983). From the late 1920s to the late 1970s, more than three-fourths of the city's solid waste sites (incinerators and landfills) were located in black neighborhoods, although African Americans made up only one-fourth of Houston's population.

In 1979 residents of the city's Northwood Manor neighborhood (where African Americans made up more than 84% of the total residents) chose to challenge the Browning-Ferris Industries' (one of the world's largest waste disposal firms) selection of their area for a garbage dump. Residents from this middle-income suburban neighborhood (some 83% of the residents own their homes) mobilized soon after they discovered the construction was not an expansion of their subdivision but a municipal landfill.

Dallas, Texas

Dallas is the nation's seventh largest city, with a population of 904,078 in 1980. African Americans represent 29.4% of the city's population. The African-American population remains segregated: eight of every ten blacks live in mostly black neighborhoods. West Dallas is just one of these segregated enclaves, and African Americans make up 85% of the neighborhood residents. This low-income neighborhood dates back to the turn of the century. For more than five decades, residents were bombarded with toxic emissions from the nearby RSR Corp. lead smelter facility (Dallas Alliance Environmental Task Force, 1983). The smelter routinely pumped more than 269 tons of lead particles each year into the air. After repeated violations and citations, residents from West Dallas in 1981 mobilized to close the plant and get the lead-tainted soil removed from their neighborhood.

Institute, West Virginia

Institute is located in the Kanawha River Valley just 6 miles west of Charleston, the state capital. It is an unincorporated community where African Americans represent more than 90% of the population. The community dates back to 1891 when the state legislature selected the site for West Virginia Colored Institute (later renamed West Virginia State College). The community is also home to the Union Carbide chemical plant that was the prototype for the company's plant in Bhopal, India. The Institute plant was the only plant in the United States that manufactured the same deadly methyl isocyanate (MIC) responsible for the 1984 Bhopal disaster. A leak on 11 August 1985 at the Union Carbide Institute plant sent more than 135 local residents to the hospital. This accident heightened an already uneasy relationship that had existed for years between Union Carbide and Institute residents. People were concerned that another Bhopal incident would happen in Institute (Franklin, 1986; Slaughter, 1985). Institute residents organized themselves into a group called People Concerned about MIC to combat this toxic threat.

Alsen, Louisiana

Alsen is an unincorporated community located on the Mississippi River several miles north of Baton Rouge, Louisiana's state capital. African Americans make up 98.9% of this community, which lies at the beginning of the 85-mile industrial corridor where one-quarter of the nation's petrochemicals are produced. This corridor has been described as a "massive human experiment" and a "national sacrifice zone" (Brown, 1987, pp. 152–161). The nation's fourth largest hazardous waste landfill is located adjacent to the Alsen community. The landfill opened in the late seventies. The site represented 11.3% of the nation's remaining hazardous waste landfill capacity in 1986 (Commission for Racial Justice, 1987). Between 1980 and 1985, Rollins, the company operating the site, was cited for more than 100 state and federal toxic emission violations but did not pay any penalties. The community began organizing in late 1981 against the Rollins hazardous waste facility.

Emelle, Alabama

This Sumter County community is located in the heart of west Alabama's economically impoverished "blackbelt." African Americans make up more than 90% of the rural community's 626 residents. Emelle is home to Chemical Waste Management's "Cadillac" of hazardous waste landfills, the nation's largest hazardous waste dump (Gunter and Williams, 1986, p. 19). The site covers more than 2400 acres and represented nearly one-fourth of the nation's hazardous waste landfill capacity in 1986. The Emelle landfill opened in 1978 and is one of two hazardous waste landfills located in EPA's Region IV (i.e., the eight states in the southeastern United States). The landfill receives some of the most toxic waste in the country, including waste from cleaned-up "Superfund" sites. Public opposition began after residents discovered the new job-generating industry was not a brick factory (as was rumored) but a toxic waste dump. The initial protest against the facility, led by members of the Minority Peoples' Council, was over worker safety, rather than risks to the larger community.

Environmental Conflict Resolution

The environmental disputes explored in this analysis were seen by grassroots leaders as based on unfair treatment and as an expression of racial discrimination. These activists saw their communities sharing a disparate burden and degree of risk associated with the industrial plants. The noxious facility siting disputes were linked to earlier civil rights disputes that centered on racial disparities. Residents in Houston and Dallas, the neighborhoods in the two large cities studied, were able to inject their environmental disputes into the local political elections. Of the five communities studied, Emelle, which is a community heavily dependent upon the millions of industrial dollars pumped into the local economy annually, gave the strongest endorsement of jobs over the environment. However, even residents from this rural community were not willing to sit silent and watch their area turned into a toxic wasteland.

The opposition strategies varied somewhat across the communities. However, several common strategies were used by residents in the affected communities. For example, all five communities used protest demonstrations, petitions, and press lobbying to publicize their plight. Three of the communities (West Dallas, Alsen, and Houston's Northwood Manor neighborhood) were successful in enlisting the assistance of government agencies in their efforts to redress their environmental problem. The West Dallas community was the only one to actually convince the city and state to join them in litigation against the industrial polluter. Houston's city council, after intense pressure from the African-American community, passed a resolution opposing the controversial municipal sanitary landfill. Alsen residents convinced state environmental officials to take action against Rollins over its toxic emissions. These same three communities also filed class-action lawsuits against the industrial firms.

Who spearheaded the local opposition against the polluting industries? There is clear evidence that indigenous social action groups and their leaders held the most important and visible roles in mobilizing the opposition to industrial polluters. African-American church leaders, community improvement workers, and civil rights activists planned and initiated most of the local opposition strategies. Mainstream environmental leaders and other "outside elites" played only a minor role in mobilizing citizen opposition in these communities. The West Dallas community was able to get a government-sanctioned citizen group, the Dallas Alliance Environmental Task Force, to work on the local lead pollution problem. Institute's People Concerned about MIC was initiated by a white professor from West Virginia State College and included a broad cross-section of the community. In Emelle, African-American civil rights activists (e.g., Minority People's Council) and white environmentalists (e.g., Alabamians for a Clean Environment) have forged a loose alliance to work on the local hazardous waste problem. This is not a small point, given the history of race relations in Alabama's blackbelt.

How were the environmental disputes resolved? The disputes in the West Dallas, Alsen, and Houston neighborhoods were resolved by governmental decisions and adjudication. Bargaining and negotiation were the chief tools used to address (though not resolve) the ongoing environmental disputes in Emelle and Institute. Only West Dallas was able to force the polluting industry to completely shut down (but not dismantle and clean up the site), whereas capacity reductions were placed on the industries in Alsen, Houston, and Institute.

Litigation brought by citizens from West Dallas and Alsen resulted in out-of-court settlement agreements in favor of the plaintiffs and fines paid to governmental regulatory agencies for pollution and safety violations. In 1985 the West Dallas plaintiffs (370 children who lived in the nearby public housing project and 40 property owners) won a $20 million settlement against RSR Corp. The Alsen plaintiffs agreed to settle their dispute for an undisclosed amount (it is reported that Rollins awarded each Alsen plaintiff an average of $3000). Government officials also fined the Institute and Emelle facilities for pollution and safety violations. Citizens in Alsen and Institute were able to extract some concessions from the firms, mainly in the form of technical modifications in the plant operations, updated safety and pollution monitoring systems, and reduced emission levels.

The federal court in 1984 ruled against the Houston plaintiffs (5 years after the suit was brought and the the site opened), and the landfill was built. The Northwood Manor residents, however, were able to force the city and state to modify their requirements for siting solid waste facilities. The Houston city council passed ordinances that (1) prohibited city-owned solid waste trucks from dumping at the controversial landfill and (2) regulated the distance that future landfills could be placed near public facilities (e.g., schools, parks, and recreation playgrounds). This was not a small concession given the fact that the city has long resisted any move to institute land-use zoning (Bullard, 1987, pp. 60–75). The Texas Department of Health, which is the state permit agency for municipal landfills, also modified its regulations requiring waste disposal applicants to include socioeconomic information on census tracts contiguous to the proposed sites.

Conclusion

In their search for operating space, industries such as waste disposal and treatment facilities, heavy metals operations, and chemical plants found minority communities to be logical choices for their expansion. These communities and their leaders were seen as having a Third World view of development—that is, any development is better than no development at all. Moreover, many residents in these communities were suspicious of environmentalists, a sentiment that aligned them with the progrowth advocates.

African Americans have begun to challenge the legitimacy of environment–jobs trade-offs. They are now asking whether the costs borne by their communities are imposed to spare the larger community. Can environmental inequities (resulting from industrial facility siting decisions) be compensated? What are "acceptable" risks? Concern about inequity—the inherent imbalance between localized costs and dispersed benefits—appears to be the driving force around which African-American communities are organizing.

A great deal of overlap exists between the leadership of African-American social action groups, neighborhood associations, and community-based grassroots environmental groups that are formed to challenge local environmental problems. Preexisting institutions and their leaders play a pivotal role in the organization, planning, and mobilization stages of the opposition activities.

The Not in My Backyard (NIMBY) syndrome has trickled down to nearly all communities, including African Americans in the suburbs, impoverished ghettos, and rural blackbelt. Few residents want garbage dumps, landfills, incinerators, or other polluting industries in their backyards. The price of siting noxious facilities has skyrocketed in recent years as a result of more stringent federal regulations and the growing militancy among communities of the poor, of the working class, and of color.

African Americans and other people of color are still underrepresented in the mainstream environmental organizations at all levels. This picture must change if the U.S. national environmental movement is to provide leadership in the global environmental movement, i.e., the Third World, where most of the world's population is located. On the other hand, by embracing environmental justice issues, grassroots environmental groups have been able to make progress toward alliances with African-American communities around the toxics issue.

Changing demographics point to a more racially diverse nation. It is time for the environmental movement to diversify and reach out to the "other" United States. This does not mean the Big Ten should swallow up grassroots efforts. The 1990s offer some challenging opportunities for the environmental movement in the United States to embrace social justice and redistributive policies. African Americans and other people of color must be empowered through their own organizations to address the problems in their communities. Some positive signs indicate that the larger environmental and social justice community is beginning to take the first steps toward reducing the artificial barriers that have kept them apart.

Acknowledgment

Research for this paper was supported by a grant from Resources for the Future Small Grants Program.

References

Babcock, Richard. 1982. Houston: Unzoned, unfettered, and mostly unrepentent. *Planning* 48:21–23.

Bloom, Jack. 1987. *Class, race and the civil rights movement.* Bloomington: Indiana University Press.

Brown, Michael H. 1987. *The toxic cloud: The poisoning of America's air.* New York: Harper & Row.

Bullard, Robert D. 1983. Solid waste sites and the black Houston community. *Sociological Inquiry* 53:273–288.

———. 1984. Endangered environs: The price of unplanned growth in boomtown Houston. *California Sociologist* 7:84–102.

———. 1990. *Dumping in Dixie: Race, class, and environmental quality.* Boulder, CO: Westview.

Bullard, Robert D., and Beverly H. Wright. 1986. The politics of pollution: Implications for the black community. *Phylon* 47:71–78.

———. 1987a. Blacks and the environment. *Humboldt Journal of Social Relations* 14:165–184.

———. 1987b. Environmentalism and the politics of equity: Emergent trends in the black community. *Mid-America Review of Sociology* 12:21–37.

Buttel, Frederick, and William L. Flinn. 1978. Social class and mass environmental beliefs: A reconsideration. *Environment and Behavior* 10:433–450.

Commission for Racial Justice. 1987. *Toxic wastes and race: A national report on the racial and socioeconomic characteristics of communities with hazardous wastes sites.* New York: United Church of Christ.

Dallas Alliance Environmental Task Force. 1983, June 29. *Final Report.* Dallas, Texas: Author.

Dunlap, Riley E. 1987. Public opinion on the environment in the Reagan era: Polls, pollution, and politics revisited. *Environment* 29:6–11, 31–37.

Feagin, Joe R. 1985. The global context of metropolitan growth: Houston and the oil industry. *American Journal of Sociology* 90:1204–1230.

———. 1987. *Free enterprise city: Houston in political and economic perspective.* New Brunswick, NJ: Rutgers University Press.

Franklin, Ben A. 1986. In the shadow of the valley. *Sierra* 71:38–43.

Freudenberg, Nicholas. 1984. Citizen action for environmental health: Report of a survey of community organizations. *American Journal of Public Health* 74:444–448.

Freudenberg, N., and C. Steinsapir. 1990, February 15–20. The grass roots environmental movement: Not in our backyards. Paper presented at the annual meeting of the American Association for the Advancement of Science, New Orleans.

Gale, Richard P. 1983. The environmental movement and the left: Antagonists or allies? *Sociological Inquiry* 53:179–199.

———. 1972. From sit-in to hike-in: A comparison of the civil rights and environmental movements. *Social behavior, natural resources, and the environment,* eds. W. R. Burch, N. H. Cheek, and L. Taylor, pp. 280–305. New York: Harper & Row.

Geiser, Ken, and Gerry Waneck. 1983. PCBs and Warren County. *Science for the People* 15:13–17.

Gianessi, L. P., H. M. Peskin, and E. Wolff. 1979, May. The distributional effect of uniform air pollution policy in the U.S. *Quarterly Journal of Economics* 93:281–301.

Gottlieb, Robert, and Helen Ingram. 1988. The new environmentalists. *The Progressive* 52:14–15.

Gunter, Booth, and Mike Williams. 1986. The Cadillac of dumps. *Sierra* 71:9–22.

Hays, Samuel P. 1987. *Beauty, health, and permanence: Environmental politics in the United States, 1955–1985.* New York: Cambridge University Press.

Humphrey, Craig R., and Frederick R. Buttel. 1982. *Environment, energy and society.* Belmont, CA: Wadsworth.

Jordon, Vernon. 1980. Sins of omission. *Environmental Action* 11:26–30.

Kazis, Richard, and Richard Grossman. 1983. *Fear at work: Job blackmail, labor, and the environment.* New York: The Pilgrim Press.

Kruvant, W. J. 1975. People, energy and pollution. In *The American energy consumer,* eds. D. Newman and D. Dawn, pp. 125–167. Cambridge, MA: Ballinger.

Levine, Adeline. 1982. *Love Canal: Science, politics, and people.* Lexington, MA: Lexington Books.

Logan, John R., and Harvey Molotch. 1987. *Urban futures: The political economy of place.* Berkeley, CA: University of California Press.

McCaull, Julian. 1975. Discriminatory air pollution: If the poor don't breathe. *Environment* 19:26–32.

Mitchell, Robert Cameron. 1979. Silent Spring/solid majorities. *Public Opinion* 2:16–20.

Mohai, Paul. 1985. Public concern and elite involvement in environmental conservation. *Social Science Quarterly* 66:820–838.

Morrison, Denton E. 1980. The soft cutting edge of environmentalism: Why and how the appropriate technology notion is changing the movement. *Natural Resources Journal* 20:275–298.

——— . 1986. How and why environmental consciousness has trickled down. In *Distributional conflict in environmental resource policy,* eds. A. Schnaiberg, N. Watts, and K. Zimmerman, pp. 187–220. New York: St. Martin's Press.

Pollack, Sue, and Joann Grozuczak. 1984. *Reagan, toxics and minorities.* Washington, DC: Urban Environment Conference, Inc.

Schnaiberg, Allan. 1980. *The environment: From surplus to scarcity.* New York: Oxford University Press.

——— . 1983. Redistributive goals versus distributive politics: Social equity limits in environmentalism and appropriate technology movements. *Sociological Inquiry* 53:200–219.

Slaughter, Jane. 1985. Valley of the shadow of death. *The Progressive* 49:50.

Taylor, Dorceta A. 1989. Blacks and the environment: Toward an explanation of the concern and action gap between blacks and whites. *Environment and Behavior* 21:175–205.

Taylor, Ronald A. 1982. Do environmentalists care about poor people? *U.S. News and World Report* 96(April):51–52.

Truax, Hawley. 1990. Beyond white environmentalism: Minorities and the environment. *Environmental Action* 21:19–30.

U.S. General Accounting Office. 1983. *Siting of hazardous waste landfills and their correlation with racial and economic status of surrounding communities.* Washington, DC: Government Printing Office.

Urban Environment Conference, Inc. 1985. *Taking back our health: An institute on surviving the threat to minority communities.* Washington, DC: Urban Environment, Inc.

Van Liere, Kent, and Riley E. Dunlap. 1980. The social bases of environmental concern: A review of hypotheses, explanations, and empirical evidence. *Public Opinion Quarterly* 44:181–197.

Deep Ecology and Radical Environmentalism

BILL DEVALL

Department of Sociology
Humboldt State University
Arcata, CA 95521
USA

Abstract *The principles of deep ecology and its philosophical critique of reform environmentalism are reviewed, and the relationship between deep ecology and radical environmentalism is described. Major radical environmental groups such as Earth First!, Greenpeace, Sea Shepherd Society, and the Rainforest Action Network are analyzed in terms of their organization, strategies, and tactics. The last section includes a discussion of the mounting environmental "counter-movement" and possible changes in the radical environmental movement.*

Keywords Deep ecology, ecotage, green politics, radical environmentalism, reform environmentalism.

The U.S. conservation and environmental movement is one of the most enduring and vital social movements of the past century. Since the founding of the Sierra Club by John Muir and his associates in 1892, the movement has persisted in various versions, despite continual differences between its conservation and preservation wings (Fox, 1985). The Sierra Club, National Wildlife Federation, National Audubon Society, and Wilderness Society are among the organizations that by 1960 had national memberships, professional staff, and highly developed institutional mechanisms for making decisions and working with government agencies, the courts, and Congress in attempting to achieve their goal of conserving natural resources. Despite significant success in helping shape public policy, these conservation-oriented organizations remained relatively small in membership. They also remained split on philosophy and tactics, although they tended to share a resource management ideology that was critiqued by a few ecologically oriented intellectuals (see, e.g., Fox, 1985).

During the 1960s and early 1970s a new wave of environmental activism and policy issues attracted attention. These issues included the effects of the explosive growth of human population; the effects of toxic wastes and pollution on air, water, and soil as well as on human health and well-being; and deforestation and human-caused extinction of other species. Established conservation organizations such as the Sierra Club and newly established environmental organizations such as the Natural Resources Defense Council grew substantially in membership during the 1970s (Mitchell, 1989). However, during the 1970s, critics within the environmental movement saw these major national organizations as less and less responsive to grassroots demands for more rapid change in public policy, too bureaucratized and centralized, and too "shallow" in ideology. The major environmental groups were also criticized for their willingness to settle for reforms in government policy without changes in our society's basic culture, including the myths of economic growth, progress, belief that technology will save us from environmental problems, and humanism.

Deep ecology was a label put forward in the early 1970s for a philosophical tendency that provided both a critique of reform or shallow environmentalism and a critique of industrial society and the anthropocentric bias of that type of society. In the following pages, the major tenets of deep ecology and the relationship between deep ecology as philosophical perspective and the emerging radical wing of the environmental movement are discussed. Then, several radical groups and movements, including Earth First!, are described, providing an overview of radical environmentalism. Finally, relations among radical and reform environmental movement organizations, the anti-environmental counter-movement, and government agencies are discussed.

Deep Ecology

The terms *deep ecology* and *deep, long-range ecology movement* were originated by Norwegian philosopher and social activist Arne Naess. In a 1973 article, Naess asserted that shallow and deep ecology can be seen as two aspects of the environmental movement. He defined *shallow ecology* as the "fight against pollution and resource depletion. Central objective: The health and affluence of people in the developed countries" (Naess, 1973, p. 97). He defined *deep ecology* as a normative, ecophilosophical movement that is inspired and fortified in part by our experience as humans in nature and in part by ecological knowledge. The literature on the deep–shallow distinction and on the historical and philosophical antecedents of the deep ecology movement has been extensively developed since 1973 (see, e.g., Fox, 1990; Sessions, 1987).

The most distinctive aspect of deep ecology is the idea of *ecocentric identification,* which Naess calls the "ultimate norm" of self-realization. Humans are one of myriad self-realizing beings, and human maturity and self-realization come from broader and wider self-identification. Out of identification with forests, rivers, deserts, or mountains comes a kind of solidarity: "I am the rainforest" or "I am speaking for this mountain because it is a part of me."

Naess says that Rachel Carson (*Silent spring,* 1962) showed this kind of self-identification combined with ecological understanding. According to Naess,

> The intense motivation of Rachel Carson was of a deep kind. It had to do with how she intuitively conceived herself, the human existence and the world. In short, a combined religious and philosophical motivation. She said: We cannot treat creation, God's creation, as we do. And it is in no way good for ourselves to behave as we do. Her Self was wide and deep. Her main motivation was religious and philosophical, *not* narrowly utilitarian. (Naess, 1989a, p. 7)

A second ultimate norm of deep ecology is *biocentrism,* or *ecocentrism* as some call it. In contrast with an anthropocentric or human-centered worldview, an ecocentric worldview suggests that humans are part of the "web of life"—not at the top of creation but equal with the many other aspects of creation. Naess calls this a *total-field image* and suggests that this image encourages respect for natural biodiversity and evolution (Naess, 1973). Speaking as a philosopher, Naess suggests that in an anthropocentric worldview every action is undertaken to protect present and future generations of humans. In an ecocentric worldview, future generations include generations of *all* living beings and "beings" are broadly defined to include living rivers as well as living species.

The supporters of the deep ecology movement, then, seek ways of living that are best for all living beings. "Do as little harm as possible" might be a slogan of those seeking a deep ecology–based lifestyle. Some taking of life is necessary to satisfy vital human needs, but the integrity, beauty, and stability of the native landscape is respected.

Realizing the goal of an ecocentric-based society requires *ecosophy* or wisdom. Ecosophy literally means wisdom of the household, or the place within which we dwell. Alan Drengson, editor of *The Trumpeter,* a widely read journal on deep ecology, explains that:

> "Ecosophy" for humans means ecological wisdom, which is not just discursive knowing, but a state of harmonious relationship with Nature. The Earth has wisdom, and each species has a wisdom peculiar to it, which is exemplified in continued flourishing, beyond bare survival. "Flourish" means an optimal state of self-fulfillment and self-realization. . . . Ecosophy is the wisdom of dwelling in a place and it is also the wisdom to dwell in a place harmoniously. (Drengson, 1990, pp. 101–102)

There could be many articulations of a deep ecology type of position. Some of these could be based on Native American traditions, on Buddhist traditions, or on other cultural traditions and philosophical perspectives. Naess himself was greatly influenced by the philosopher Spinoza. Naess calls his own articulation of deep ecology "Ecosophy T" (Naess, 1989b).

Sensing that some kind of platform or general statement was needed to show the unity among a diversity of deep ecology types of positions, Naess, along with philosopher George Sessions, formulated "8 points" as a modest suggestion for discussion. Naess insists that this platform is without great pretensions and has the primary function of stimulating dialogue about philosophy and strategies in politics and personal lifestyle decisions (Devall and Sessions, 1985). Naess and other supporters of deep ecology have expanded and extensively commented on these principles in many publications (see Fox, 1990).

(1) The well-being and flourishing of human and nonhuman life on Earth have value in themselves. These values are independent of the usefulness of the nonhuman world for human purposes.
(2) Richness and diversity of life forms contribute to the realization of these values and are also values in themselves.
(3) Humans have no right to reduce this richness and diversity except to satisfy human needs.
(4) The flourishing of human life and cultures is compatible with a substantial decrease of the human population. The flourishing of nonhuman life requires such a decrease.
(5) Present human interference with the nonhuman world is excessive, and the situation is rapidly worsening.

(6) Policies must therefore be changed. The changes in policies affect basic economic, technological, and ideological structures. The resulting state of affairs will be deeply different from the present.
(7) The ideological change is mainly that of appreciating life quality (dwelling in situations of inherent worth) rather than adhering to an increasingly higher standard of living. There will be a profound awareness of the difference between big and great.
(8) Those who subscribe to the foregoing points have an obligation directly or indirectly to participate in the attempt to implement the necessary changes. (Devall and Sessions, 1985, p. 70)

Although a wide variety of lifestyles and social policies are potentially compatible with a deep ecology position, the literature on deep ecology suggests that many supporters favor what has been called *green consumerism* (careful awareness both of the quality and quantity of products consumed, based on the principle of least harm to living beings and ecocentric identity), voluntary simplicity of lifestyle that maximizes rich experiences in nature (Devall, 1988), and bioregionalism or living in place (Sale, 1985). Supporters of deep ecology also tend to encourage the restoration movement, which seeks to enhance and restore native biodiversity within a bioregional context (see Berger, 1985), and to favor protection of ancient forests, tropical rainforests, and all other types of ecosystems on the planet. Some supporters of deep ecology also favor vegetarianism based on principles outlined in John Robbins's (1987) *Diet for a new America.* Robbins emphasizes eating lower on the food chain. Generally speaking, the norm of nonviolence is widely accepted by deep ecologists. Naess himself wrote an explanation of Gandhi's principles of nonviolence and has been interested in Gandhian types of social movements since the 1930s (see Naess, 1974).

Deep Ecology and Green Politics

Supporters of deep ecology share with "greens" and green socialists a criticism of modern industrial societies and of the failure of reform environmental groups to make fundamental critical assessments of industrial society (e.g., Faber and O'Connor, 1989). Deep ecologists tend to differ from green socialists as well as from ecofeminists (Plant, 1989) in assessing the reasons for the current environmental crisis. Although green socialists look to the conditions of capitalism and ecofeminist theorists focus on androcentrism as the cause of the environmental crisis, deep ecologists emphasize the dangerous tendencies of anthropocentrism, including the tendency toward hubris and narrow conceptions of the ego as a power-seeking entity. Deep ecologists also suggest that a rich life means exploring our "wild self" or ecological self and suggest that to do this we need to unmask the anthropocentrism that is part of the basic ideology of industrial society.

Warwick Fox, responding to critics of deep ecology, contrasts the deep ecology position with positions of other radical critical theories (Fox, 1989). Deep ecology has strong affinities with established green parties and with emerging green political movements, as can be seen by comparing the deep ecology platform listed earlier with the principles and criteria of green parties and green activism (see, e.g., Capra and Spretnak, 1986; Porritt, 1985). The four pillars of green philosophy—ecology, grassroots democracy, social responsibility, and nonviolence—are variously interpreted, and healthy dialogue has occurred. The most revolutionary of the pillars, however, is ecology. Moreover, as Robyn Eckersley points out, the most radical and promising ecological insights have emerged in North America and Australasia rather than in Western Europe where green parties are most fully established (Eckersley, 1990).

Although supporters of deep ecology acknowledge the importance of developing compatible strategies for social justice and ecology, some green theorists have yet to accept that nature has inherent worth or even the importance of protecting ecosystem integrity—such as tropical rainforests—to enhance human well-being on the planet. Many supporters of deep ecology reject in turn any political strategy that ties economic growth to invasion of any remaining habitat of endangered species or massive developments of rainforests or other areas that provide habitat for native species. The uneasiness that many green socialists and social ecologists feel with deep ecology and the vitriolic criticisms that some green socialists and social ecologists, including Murray Bookchin, have heaped on it reflect their unwillingness to accept its rejection of anthropocentrism for a truly ecocentric perspective (Manes, 1990).

Deep Ecology and Radical Environmentalism

As noted earlier, the deep, long-range ecology movement is a philosophical movement with implications for personal lifestyles and public policy as suggested in the platform presented by Naess and Sessions. Yet, deep ecology has increasingly become associated with radical environmental activism, or the use of tactics such as ecotage, sit-ins, guerrilla theater, demonstrations, and other forms of direct action (Manes, 1990). Although deep ecology does not provide a formal ideology for radical environmentalism, it offers a "diverse body of ideas . . . which taken as a whole express the vision behind the activism" (Manes, 1990, p. 136). Part of this vision consists of a rejection of the shallow perspective and reformist tactics of the mainstream environmental movement.

Many radical environmentalists believe that, despite their anthropocentric orientation, reform environmental groups such as the Sierra Club, National Audubon Society, and Wilderness Society had a "window of opportunity" during the past few decades to work through conventional political channels for necessary changes. As noted by Peter Borrelli, editor of *The Amicus Journal,* particularly during the 1980s, many activists became increasingly disenchanted and alienated from mainstream environmental organizations. They were discouraged by the compromising attitude of mainstream groups, by the bureaucratization of the groups, by the professionalization of leaders and their detachment from the emerging concerns of grassroots supporters, and by the lack of success of mainstream organizations in countering the Reagan anti-environmental agenda (Borrelli, 1988).

The emergence of so-called third-wave environmentalism during the 1980s especially discouraged and alienated many environmentalists. Third-wave environmentalism was based on the principle that environmental experts, usually lawyers and scientists, could and should negotiate directly with corporations and government agencies to achieve compromises on pollution controls, energy policies, and other environmental issues, preferably using the "market" mechanism. Third-wave environmentalism is narrowly rational and focuses on economics and public policy. No recognition is given to ecological sensibilities, the necessity of providing an ecocentric critique of industrial society, or of exploring the "wild self" and transforming society. Grassroots activists were not consulted in setting policy and, indeed, grassroots activists were frequently seen as burdened with too much emotion. The Environmental Defense Fund is sometimes cited as a prototypical third-wave group (see Borrelli, 1988).

Christopher Manes (1990), in his book *Green rage,* summarizes how reform environmentalism stimulated the development of radical environmentalism:

> The impetus for the radical environmental movement, at least in this country (USA), was not solely a response to the smug advocates of wilderness destruction and industrial development. . . . The mainstream, reformist environmental movement, embodied in national groups like the Sierra Club and the Wilderness Society, ensured an explosion of radical environmental forces by anxiously trying to restrain them. (Manes, 1990, p. 44)

Led by the example of people like Dave Foreman and Howie Wolke, cofounders of Earth First! (both ex-staff members of mainstream environmental groups), a growing number of grassroots activists drew inspiration from the civil rights, anti-Vietnam, and women's movements and explored various forms of direct action—civil disobedience, guerilla theater, monkey-wrenching, nonviolent demonstrations, and anarchism—in their efforts to open the minds and hearts of their fellow citizens to the plight of the planet under the domination of industrial society. In doing so they were continuing the radical amateur tradition that Fox (1985) sees as having been crucial throughout the history of the conservation movement.

Radical environmentalists are also determined to reclaim their spiritual identity with nature, whether in the Buddhist tradition espoused by Gary Synder and Robert Aitken, in the tribal rituals of the early years of Earth First!, or in the "Council of All Beings" ritual (Seed, Macy, Fleming, and Naess, 1988). As described by John Seed, the purpose of a Council of All Beings is to change the perception of humans as *above* nature to that of humans *in* nature:

> What is described here should not be seen as merely intellectual. The intellect is one entry point. . . . For some people, however, this change of perspective follows from actions on behalf of mother Earth. "I am protecting the rain forest" develops to "I am part of the rain forest protecting myself. I am that part of the rain forest recently emerged into thinking." What a relief then! The thousands of years of imagined separation are over and we begin to recall our true nature. That is, the change is a spiritual one, thinking like a mountain, sometimes referred to as "deep ecology." (quoted in Devall and Sessions, 1985, p. 243)

Although radical environmentalism has been stimulated by the failures of reform environmentalism and by the philosophy and spirituality of deep ecology, ultimately "it is based on one simple but frightening realization: that our culture is lethal to the ecology that it depends on" (Manes, 1990, p. 22). At root, then, radical environmentalism is a response to our existential condition. In a culture dominated by humanism and technology, radical environmental sensibilities come from a sense of the peril that all beings face because of human intervention in the biosphere. The agenda of radical environmentalism is less and less a reaction to the agenda of reform environmentalism (or the activities of industry and government) and more and more a reaction to the demands of our existence. Its concerns are acid rain, increasing rates of species extinction, the greenhouse effect, ozone depletion, and on and on.

Radical Environmental Movements

Supporters of radical environmentalism and deep ecology are found in many organizations. Indeed, for many radical environmentalists, organizations are simply expedient arenas for action. Even founders of radical organizations are willing to abandon those organizations when the demands by conservative members to maintain the organization (''organizational maintenance problems'' in sociological terminology) threaten to overwhelm the ideological purpose of the group. Thus Dave Foreman, one of the cofounders of Earth First!, was willing to abandon Earth First! in the early 1990s and move on to form another project in defense of native biodiversity after Earth First! had served its purpose of making mainstream environmental groups look moderate.

Radical environmentalists work in affinity groups protecting specific ecosystems from the attacks of industrial society, or work in green movements, in ad hoc groups organizing protests and demonstrations, and in environmental education groups. In this section several of the major organizations are discussed, beginning with Earth First!

The Earth First! movement began in 1980. A group of friends, several of whom were staff members of mainstream environmental groups, were returning from Mexico where they had discussed their dissatisfaction with the compromising attitudes of their employers. They began shouting the slogan ''Earth First!'' By the time they got back to the Arizona border, they were determined to start a movement to mobilize support for big wilderness, for the defense of biodiversity (Manes, 1990). The founders of Earth First! were inspired by the writings of Edward Abbey, especially his novel *The monkey wrench gang* (1975). It was fitting, then, that Abbey attended a ceremony on 21 March 1981 at the Glen Canyon Dam, where the dam was ''cracked'' by monkey-wrenching Earth First!ers in one of their first ''actions.'' In some ways, the Glen Canyon Dam is the quintessential symbol of industrial society run amok and the drowning of wild nature. ''Cracking'' the dam, therefore, symbolically disempowered the juggernaut of industrial society (Manes, 1990).

Starting in the U.S. Southwest under the slogan ''rednecks for wilderness,'' Earth First! spread across the continent and on to Australasia and Europe. Although not committed to any specific political tactics, speaking in many voices, and disavowing bureaucracy, centralized decision making, sexism, and hierarchy, Earth First! was a vortex of action in radical environmentalism during the 1980s (Manes, 1990). It was variously described as anarchists, a tribe, a collection of social deviants, ecoterrorists, and visionaries. Some Earth Firster!ers proposed defending wilderness by monkey-wrenching logging equipment and oil exploration equipment and by other acts of ''ecotage.'' Although emphasizing that he did not speak for the movement, Earth First! Dave Foreman's book *Ecodefense* (1988) has been considered as a definitive manual for ethical monkey-wrenching.[1] In a 1984 interview, Edward Abbey said that in *The monkey wrench gang,*

> [I] tried to make a clear distinction between sabotage and terrorism. My 'monkeywrenchers' were saboteurs, not terrorists. Sabotage is violence against inanimate objects: machinery and property. Terrorism is violence against human beings. I am definitely opposed to terrorism, whether practiced by the military and state—as it usually is—or by what we might call unlicensed individuals. (Abbey, 1984, p. 18)

''Do not harm human lives'' has always been a principle of Earth First! and other radical organizations.

The movement Earth First!, however, is much more than monkey-wrenching. Foreman and his associates have developed wilderness inventories and innovative suggestions for preserving biodiversity. *The big outside,* compiled by Foreman and Wolke (1989), is the most extensive inventory and description of big wilderness in the United States ever undertaken by private individuals. The defense of big wilderness and native biodiversity for its inherent worth became the central focus of radical environmentalism during the 1980s and defined the difference between radical environmentalism and other brands of radicalism including urban anarchism, social ecology, and feminism.[2]

While Earth First!ers were working as "ecowarriors" on land, Greenpeace was working to protect the atmosphere from the testing of nuclear weapons and defending the whales in the ocean. Greenpeace was founded in Vancouver, British Columbia, in 1971 by "radical amateurs" who understood the threat of atmospheric testing of nuclear weapons (Hunter, 1979). When Greenpeace took up the whaling issue in the 1970s, they created the first international radical environmental movement. As Day concludes, commercial whaling, by all rational economic accounting, should have ended by 1970. Only the encouragement of some government agencies and the hubris of specific men keep the commercial whaling fleets on the sea (Day, 1987). Greenpeace volunteers aroused worldwide interest in the plight of the whales and counteracted the worldwide effort of the Japanese government to keep commercial whaling fleets on the ocean.

Greenpeace was also possibly the first environmental group to have government terrorist action directed against one of its projects. Agents of the French secret service bombed a Greenpeace vessel in Auckland Harbor, New Zealand, in 1985, killing a photographer on board (Day, 1989).

Although nonviolence is a central norm for radical environmentalism, there are various interpretations of this norm. The Sea Shepherd Society was founded by Paul Watson after he was expelled from Greenpeace over an interpretation of nonviolence. Watson mobilized volunteers to oppose the killing of baby seals off the coast of Canada and to oppose illegal whale hunting in the North Atlantic. In one of his most publicized actions he used his own boat to ram a whaling boat off the coast of Portugal (Watson and Rogers, 1982). In another highly publicized action of volunteers within the Sea Shepherd Society, two men flew to Iceland, one of the last whaling nations, and opened the sea hatches of two Icelandic whaling ships, allowing them to sink to the bottom of the harbor. Paul Watson took responsibility for organizing this action and later flew to Iceland to talk with government officials. After detaining him, the officials released him and sent him back to Canada (Dykstra, 1986).

Rainforest Action Network (RAN; office in San Francisco) and Rainforest Information Center (RIC; office in Lismore, Australia) exemplify a new type of radical environmental organization. They call themselves "networks" that bring together people interested in one topic, in these cases the plight of rainforests. The purpose of a network is to widely disseminate information and to empower grassroots activists to protest specific actions of corporations and government agencies. For example, RAN and RIC use extensive computer networks such as ECONET to inform members of actions by corporations. Furthermore, frequent "action alerts" are sent to people on their mailing lists asking for letters of support for rainforest activists. Randy Hayes, founder of RAN, and John Seed, founder of RIC, are both supporters of deep ecology. Both RAN and RIC emphasize the importance of protecting native biodiversity and the human rights of native tribal groups who have inhabited rainforests for thousands of years. Both present information on the assaults by governments and multinational corporations on the integrity of rainforests around the Earth. Both use diverse tactics, including letter writing campaigns, demonstrations at the annual meetings of the World Bank and stockholders' meetings of multinational corporations, boycotts of selected corporations, appeals to governments, and varied educational campaigns directed at different segments of the population in North America, Europe, and Australasia.[3] Both appeal to the governments of Japan, Canada, and the United States, and the Common Market of Europe, the primary players in the world tropical rainforest deforestation game.

Both RAN and RIC rely on what writer Richard Grossman calls "bearing witness." "Bearing witness" comes from practices developed by the Society of Friends and other nonviolent Christian and Buddhist traditions. Bearing witness means to stand forth or to be an example for others. For radical environmentalists exploring their own spiritual awakening in an age of environmental crisis, this means bearing witness for the mute forests that are being burned and chainsawed in the name of progress, greed, or ignorance (Grossman, 1988). It is perhaps the strongest link between radical environmentalism and spiritually based deep ecology and green politics (Devall and Sessions, 1985; Spretnak, 1986).

Radical environmentalism is clearly developing a diverse organizational base (ranging from the local to the international level), although one that is far more informal and decentralized than that of the mainstream environmental organizations. Radical environmentalism, by using networks, affinity groups, and other organizational techniques, may eventually develop into a full-fledged alternative to the multi-faceted mainstream environmental movement.

Trends in Radical Environmentalism

In April 1990, millions of Americans, Canadians, and residents of many other nations joined together to celebrate the twentieth anniversary of Earth Day. *Time* magazine, in 1989, had already declared Earth the "Planet of the Year," instead of naming their usual man or woman of the year. "Earth: The endangered planet" was headlined in numerous magazines, newspapers, and television reports. On Earth Day 1990, major industries such as the U.S. cattle industry and the oil and gas industry, as well as timber trade associations, took out full-page ads in newspapers, proclaiming "Every day is Earth Day for us." George Bush proclaimed during his 1988 presidential campaign, "I am an environmentalist." Earth Day 1990 was celebrated with concerts and rhetoric, and green consumerism was heralded as the salvation of the environment.

Less than a month later, on 24 May 1990, a pipe bomb exploded in a car carrying two Earth First! radical environmentalists in Oakland, California. Darryl Cherney and Judi Bari were organizing a "Redwood Summer" to protest the rapid clear-cutting of ancient forests in northern California (Carothers, 1990). Charging that police had mishandled the investigation of the bombing, indeed initially arresting Bari and Cherney on charges of transporting the bomb, a coalition of environmental groups and Congresspersons in July 1990 called for a Congressional investigation of police treatment of these Earth First! activists.

In July 1990, Secretary of Interior Lujan and President Bush were calling for changes in the Endangered Species Act to "balance" economic growth and species preservation. At the conclusion of the Economic Summit in Houston in July 1990, President Bush lashed out at "extreme" environmentalists. Leaders of mainstream environmental organizations had held a countersummit in Houston, asking the seven industrial nations to change their policies to reduce the so-called greenhouse effect. Meanwhile, Secretary Lujan was testifying before a Congressional committee on a Bush administration request to allow offshore oil leasing.

While reform environmental organizations continue to appeal to an administration that has turned a deaf ear to nature, radical environmentalists indict industrial civilization itself. A key issue affecting the future of the environmental movement will probably be the relationship between the new radical wing of the movement and the mainstream environmental organizations such as the Sierra Club. The Sierra Club refused to participate in demonstrations during Redwood Summer, reportedly because of fears relating to their liability insurance coverage. It is still problematic whether actions such as the Redwood Summer will prove to be a valuable resource to mainstream environmental groups by making them appear more moderate, as sometimes suggested. As Redwood Summer continued in Humboldt County, California, even the most moderate environmental groups were attacked. Protesters even marched in front of the Saturday food market protesting organic food. "These attacks by environmentalists on agriculture and forest industries must stop," protesters said. Carrying signs saying "Take back America," "America, love it or leave it," "Earth First out of Humboldt County," and "No more lawsuits" (referring to the many lawsuits in which timber corporations were found by the courts to be in violation of the law), some timber workers and their spouses denounced attempts to enforce environmental laws in California.

The countermovement of anti-environmentalism is in full swing as of this writing.[5] Claiming that environmentalists are pushing "the big lie" and that "there is no problem with the timber industry or with agriculture," timber worker groups in the Pacific Northwest during the summer of 1990 were proclaiming that environmentalism is "anti-American," thus echoing President Bush's statement that he is taking a "reasonable" position on environmental issues and that environmentalists who seek greater protection of endangered species are "extremists." An example of the counter-movement tactics is seen in a speech given by Bill Holmes, a former member of the California Board of Forestry, to the Redwood Region Logging Conference, 15 March 1991. He called for a "People First!" movement with a single objective—to expose environmentalism. He went on to say, ". . . I'm betting that there are some of you out there who are tired of turning the other cheek and are ready to kick somebody in the crotch" (Holmes, 1991, p. 50).

The FBI has engaged in extensive investigations of some radical environmentalists, including undercover operations against Arizona Earth First! and investigations of Greenpeace (Manes, 1990). David Day, in *The environmental wars* (1989), documents the murders and deaths of environmental martyrs, including Chico Mendez.

Although federal and state governments are increasing their police forces in national forests in response to what they claim are tree spiking incidents and although some local district attorneys are seeking stiffer penalties against persons found guilty of chaining themselves to the gates of lumber mills and trespassing on private land to protest aerial spraying of herbicides, mainstream environmental groups seem ossified and unable to respond to attacks on their own traditional weapons (Carothers, 1990). The rights of citizen groups, for example, to challenge timber harvest plans on federal lands in the Pacific Northwest were limited by Congress in 1989. Such counterattacks will probably continue as the push for environmental protection increases.

Furthermore, given that mainstream environmentalism and mainstream groups have failed to foster the development of any type of "humans-in-nature" spirituality and have rarely fostered creative expressions of humans-in-nature such as poetry, art, music, new interpretations of ancient myths, or rituals bonding humans to nature, it is likely that people who yearn for such expressions will turn increasingly to the deep ecology movement and to radical environmentalism. Intellectual articulation of deep ecology will continue to develop among a network of philosophers, social scientists, historians, and activists through publications such as *The Trumpeter* and *Environmental Ethics*. Within the broad framework of providing an ecological critique of industrial civilization and its anthropocentric thinking, deep ecology will no doubt continue to evolve as it absorbs important elements of transpersonal psychology, Eastern philosophies, and ecofeminist theory, as well as the science of conservation biology and a growing concern with the welfare of native peoples.

The ecocentric worldview and strong spiritual identification with the natural world represented by many people in the deep ecology movement promises to provide a potent source of inspiration for radical environmentalism and a challenge to the mainstream environmental movement as well as the rest of industrial civilization. Its ability to learn from earlier critiques by feminists and to absorb the insights of ecofeminism reflect the evolutionary strength of deep ecology (Fox, 1989). Participants in the deep, long-range ecology movement continue to learn and grow in a search for new meanings; to develop, explore, and integrate their embeddedness in nature; and to search for lifestyles and public policies that are less exploitative of nature than the conventional lifestyles of many people in the middle and upper-middle classes in advanced industrial societies. In interaction with supporters of other forms of environmental philosophy, Arne Naess and other supporters of deep ecology continually emphasize the importance of working with and finding common ground with those who speak from these different perspectives, an intellectual endeavor akin to the martial arts of Aikido.

Notes

1. Although "ecotage" was practiced by a small number of environmentalists in earlier years (see, e.g., Love and Obst, 1972), it received limited attention before the emergence of Earth First!

2. Research results from the emerging field of conservation biology have supported the argument for the need for big wilderness (Soule, 1985).

3. For example, RAN organized the boycott of Burger King in 1984 to protest deforestation of Latin American land for beef production and to educate consumers about the connection between beef production, deforestation, and domination of elites in Latin America.

4. See Gale (1986) for a theoretical discussion of the relationship between environmental movement organizations, counter-movement organizations, and government agencies.

References

Abbey, E. 1975. *The monkey wrench gang.* New York: J. B. Lippincott.

———. 1984. The plowboy interview: Slowing the industrialization of Planet Earth. *Mother Earth News,* May/June:17–24.

Berger, J. J. 1985. *Restoring the Earth: How Americans are working to renew our damaged environment.* New York: Knopf.

Borrelli, P., ed. 1988. *Crossroads: Environmental priorities for the future.* Washington, DC: Island Press.

Capra, F., and C. Spretnak. 1986. *Green politics,* rev. ed. Santa Fe, NM: Bear and Company.

Carothers, A. 1990. The new pitch of battle. *E: The Environmental Magazine,* 1(September/October):80.

Carson, R. 1962. *Silent spring.* Boston: Houghton Mifflin.

Day, D. 1987. *The whale war.* San Francisco: Sierra Club Books.

———. 1989. *The environmental wars: Reports from the front lines.* New York: St. Martin's Press.

Devall, B. 1988. *Simple in means, rich in ends: Practicing deep ecology.* Salt Lake City, UT: Peregrine Smith.

Devall, B., and G. Sessions. 1985. *Deep ecology.* Salt Lake City, UT: Peregrine Smith.

Drengson, A. 1990. In praise of ecosophy. *Trumpeter,* 7(Spring):101–103.

Dykstra, P. 1986. Greenpeace. *Environment* 28(July), p. 5.

Eckersley, R. 1990. Green theory and practice in old and new worlds: A comparative perspective. Unpublished manuscript, University of Tasmania.

Faber, D., and J. O'Connor. 1989. The struggle for nature: Environmental crisis and the crisis of environmentalism in the United States. *Capitalism Nature Socialism,* 2(Summer):12–39.

Foreman, D. 1988. *Ecodefense,* rev. ed. Tucson, AZ: Ned Ludd Books.

Foreman, D., and H. Wolke. 1989. *The big outside; A descriptive inventory of the big wilderness areas of the U.S.* Tucson, AZ: Ned Ludd Books.

Fox, S. 1985. *John Muir and his legacy: The American conservation movement.* Madison: University of Wisconsin Press.

Fox, W. 1989. The deep ecology–ecofeminism debate and its parallels. *Environmental Ethics,* 11:5–25.

———. 1990. *Toward a transpersonal ecology.* Boston: Shambala.

Gale, R. P. 1986. Social movements and the state: The environmental movement, countermovement, and government agencies. *Sociological Perspectives,* 29:202–240.

Grossman, R. 1988. *And on the eighth day we bulldozed it.* San Francisco: Rainforest Action Network.

Holmes, B. 1991. Weirdos, wimps, and watermelons. *Earth Island Journal* (Summer), p. 48.

Hunter, R. 1979. *Warriors of the rainbow: A chronicle of the Greenpeace movement.* New York: Holt, Rinehart and Winston.

Love, S., and D. Obst, eds. 1972. *Ecotage.* New York: Bantam Books.

Manes, C. 1990. *Green rage: Radical environmentalism and the unmaking of civilization.* Boston: Little, Brown.

Mitchell, R. C. 1989. From conservation to environmental movement: The development of the modern environmental lobbies. In *Government and environmental politics,* ed. M. T. Lacey. Washington, DC: Wilson Center.

Naess, A. 1973. The shallow and the deep, long-range ecology movement. *Inquiry,* 16:95–100.

———. 1974. *Gandhi and group conflict: An exploration of Satyagraha. Theoretical background.* Oslo, Norway: Universitetsforlaget.

———. 1989a. *Ecology and ethics.* Unpublished manuscript, University of Oslo.

———. 1989b. *Ecology, community, and lifestyle,* trans. and ed. D. Rothenberg. New York: Cambridge University Press.

Plant, J. ed. 1989. *Healing the wounds: The promise of ecofeminism.* Santa Cruz, CA: New Society Publishers.

Porritt, J. 1985. *Seeing green: The politics of ecology explained.* New York: Basil Blackwell.

Robbins, J. 1987. *Diet for a new America.* Walpole, NH: Stillpoint.

Sale, K. 1985. *Dwellers in the land: The bioregional vision.* San Francisco: Sierra Club Books.

Seed, J., J. Macy, P. Fleming, and A. Naess. 1988. *Thinking like a mountain: Toward a Council of All Beings.* Santa Cruz, CA: New Society Publishers.

Sessions, G. 1987. The deep ecology movement: A review. *Environmental Review,* 11:105–125.

Soule, M. E. 1985. What is conservation biology? *BioScience,* 35:727–734.

Spretnak, C. 1986. *The spiritual dimensions of green politics.* Santa Fe, NM: Bear and Company.

Watson, P., and W. Rogers. 1981. *Sea shepherd: One man's crusade for whales and seals.* New York: Norton.

Globalizing Environmentalism: Threshold of a New Phase in International Relations

LYNTON K. CALDWELL

School of Public and Environmental Affairs
Indiana University
Bloomington, IN 47405
USA

Abstract *Sociologist Robert Nisbet conjectured that "when the history of the twentieth century is finally written, the single most important social movement of the period will be judged to be environmentalism" (1982, p. 10). This assessment has been reinforced by subsequent events. Environmental concern has risen to the top of political agendas in the United States and many other countries, becoming a major consideration in international relations. A major force for the globalization of environmental politics in the United States has been the growing international concern of nongovernmental environmental organizations (NGOs). Almost all of the larger environmental NGOs now have significant international programs that attempt to influence public policies. Transnational collaboration between governments and NGOs on environmental issues has become a characteristic of contemporary politics in North America and Western Europe. Global issues such as climate change, ozone depletion, transboundary air and water pollution, endangered species, and uses of outer space will henceforth give an international dimension to national environmental policies.*

Keywords Changing perceptions, global environmental problems, international agreements, nongovernmental organizations, scientific findings, sustainable development, United Nations conferences.

As the twentieth century nears its close, environmental events have introduced new elements into the basic conditions under which nations relate to one another. "The twentieth century may, indeed, come to be seen by future generations as the time in which the concept of sovereignty and the nation state reached its apogee and began to provide our human kind with diminishing returns" (Ramphal, 1987). Events have changed traditional relationships, and obligations declared through international law have gained relevance as national boundaries cease to be effective barriers against external environmental hazards. Discovery of the risks associated with global climate change and depletion of stratospheric ozone have added new dimensions to international environmental policy.

A series of catastrophic environmental events during the 1980s—Bhopal, Sandoz, Chernobyl, Exxon Valdez—captured public attention and pushed governments to assuage popular anxiety through reassuring rhetoric and intergovernmental declarations. Action, however, has been slower to materialize, in part because the institutional structures and managerial strategies available to implement international agreements have not been adequate to the needs. Worldwide communication made possible the rapid spread of information on all issues of universal concern, and threats to the human environment are prominent among them. The activities and goals of the U.S. environmental movement were interactive with similar efforts in other countries. U.S. environmentalism was essentially moving with a worldwide tide, but it was the nongovernmental sector that was leading rather than the government in Washington. The global concerns of U.S. citizens differed in no significant respect from those of environmentalists in Canada, Western Europe, and even in the Union of Soviet Socialist Republics (U.S.S.R), but the U.S. government tended toward indifference or resistance to international environmental protection efforts. Only at the end of the 1980s did a rhetorical concession to the importance of environmental protection achieve prominence.

Legitimizing Global Environmentalism: The Legacy of Stockholm

The term *environmentalism* predates its present popular usage. It has always implied a high, sometimes inordinate, level of concern for environmental influences and relationships. Ecologically informed environmentalism has been a factor in public affairs for hardly more than two decades, although it has deeper historical roots (McCormick, 1989; Petulla, 1980). Emerging as a popular movement in the United States during the 1960s, it gained political and international legitimacy at the 1972 United Nations Conference on the Human Environment in Stockholm. During the 1960s and 1970s, the United States was foremost among nations in environmental policy and law. After 1980, the Reagan administration reversed the environmental policies of the U.S. government and thereby diminished national credibility abroad (Caldwell, 1984).

The legacy of Stockholm was to lift a popular protest movement to an agenda item in international relations, hitherto largely confined to scientifically advanced industrial states. It stimulated a worldwide awareness of environmental concerns comparable with what Earth Day 1970 did in the United States. The 1972 United Nations Conference on the Human Environment gave political legitimacy to environmental issues in international affairs (Caldwell, 1991; McCormick, 1989). Although environmental factors previously had been present in international disputes (e.g., the Trail Smelter arbitration and the Corfu incident), the points in contest related more to economic damage and to safety in navigation than to environmental policy per se (Dinwoode, 1972; Livingston, 1968). In common law countries, notably in the United States, mere environmental degradation was rarely regarded as a compensatory cause in court. Unless monetary losses could be demonstrated, environmental effects were seldom regarded as legitimate claims for damages, particularly when large numbers of people were equally affected and class-action suits were not entertained. Moreover, environmental effects were often difficult to trace with certainty to demonstrably causal sources. For example, for many years the effects of photochemical smog were not regarded as cause for compensation because the specific harmful agent in the gaseous mixture could not be linked with positive assurance to a specific human complaint.

Four developments in the mid-twentieth century converged to change the status of the environment as an object of political action and legal prescription. Without this convergence over much of the world, the Stockholm Conference would not have occurred nor its aftermath have expressed a new dimension of international relationships.

These convergent factors may be identified as (a) advances in environmental science, (b) proliferation of environment-altering technologies, (c) rise of public awareness of environmental deterioration, and (d) broadening concepts of law. These events occurred within roughly the same time period in all advanced industrialized countries. The environmental movement was inherently transnational.

Advances in Environmental Science

Although the attitude of many scientists toward early environmentalism may have been equivocal, the findings of science, notably after the International Geophysical Year (1957–1958) and the International Biological Programme (1963–1974), began to lay a factual foundation for science-based environmentalism. In 1969, while preparations for the Stockholm Conference were under way, the International Council of Scientific Unions (ICSU) established the Scientific Committee on Problems of the Environment (SCOPE). These efforts were complemented by many other multidisciplinary, multinational scientific investigations, some under ICSU sponsorship, some by specialized agencies of the United Nations (especially the World Meteorological Organization and the Educational, Scientific, and Cultural Organization), and some by nongovernmental organizations such as the International Union for Conservation of Nature and Natural Resources (IUCN) and the World Wildlife Fund (WWF).

These international collaborative efforts fed back into the agenda-setting mechanisms of national governments and intergovernmental organizations and, more important, into the perceptions of informed people and organizations at the national level. Moreover, research at the national level in atmospherics, soil science, oceanography, environmental toxicology, and especially ecology produced a rapidly expanding body of data on a wide range of environmental relationships and effects. Even in the medical profession, which for decades had tended to discount environmental influences, conferences were held to consider the "rediscovery of the environment" (Rosen, 1964). Although these scientific developments occurred primarily in technologically advanced countries, scientists in the Third World emerged as focal points for the propagation of environmental awareness in their own countries. In 1971 at the 12th Pacific Science Conference meeting in Canberra, Australia, SCOPE convened a meeting of scientists from Third World countries to consider international environmental issues. Similar meetings were held in several parts of the world in anticipation of the Stockholm Conference.

Proliferation of Environment-Altering Technologies

In the United States, by the mid-1960s the century-old love affair between modern society and technology had begun to cool. Rachel Carson's book *Silent spring* (1962) became the most widely known of a number of books reporting the adverse effects of indiscriminate uses of technology. President Johnson spoke of "the dark side to technology" and, in 1972, the U.S. Congress established the Office of Technology Assessment to test and forecast the effects of technological innovation on health, safety, and the environment. Broader in scope and inclusive of technological effects was the environmental impact statement requirement, Section 102 (2c) in the U.S. National Environmental Policy Act of 1969 (NEPA). In its 30 January 1970 issue *Life* magazine declared "Ecology: A cause becomes a mass movement" (Caldwell, 1982).

The advent of nuclear energy was a factor, among many others, in globalizing environmental policy concerns. The environmental risks associated with peaceful uses of nuclear energy were recognized internationally as early as 1957 with establishment by treaty of the International Atomic Energy Agency. Popular apprehension over the effects of radioactive fallout from atmospheric testing of nuclear weapons led to the Partial Nuclear Test Ban Treaty of 1963. The race into space by the United States and the U.S.S.R. prompted the signing of the Treaty on Principles Governing the Activities of States in the Exploration and Use of Outer Space, Including the Moon and Other Celestial Bodies, on 27 January 1967 in London, Moscow, and Washington. Transboundary radioactive contamination from an accident in 1986 at the Chernobyl nuclear reactor in the Soviet Union led to the rapid consummation of two multinational treaties for prompt notification and emergency assistance in cases of nuclear accidents.

Thus, the globalizing of policy regarding atomic and space technology, beginning prior to the Stockholm Conference, has become an unequivocal fact in international relations. Environmental effects of technologies associated with industrial chemistry, extraction and transport of petroleum, biotechnology, and uses of the electromagnetic spectrum and of outer space had become objects of concern before Stockholm. Their global significance has been subsequently reinforced by continuing incidents and international political events, notably international conferences and declarations leading to multinational programs and treaties. The mere listing of environment-related political actions occurring between 1970 and 1990 could easily fill a small book (Caldwell, 1991). These were not inadvertent happenings; they were expressions of a popular movement that began in advanced industrial countries and spread around the globe with unforeseen speed.

Rise of Public Awareness

International action to address environmental issues could not occur in the absence of national concern sufficient to place the issue on the agenda of international relations and negotiations. The practical expression of a globalized issue is its acceptance as an object of negotiation among national governments. No amount of popular concern over transboundary environmental problems—for example, over nuclear radiation, pollution of air, water, and outer space, export of hazardous materials, loss of the world's genetic heritage, or the spread of contagious disease—can lead to effective action without the involvement of government. However, governments (and private corporate organizations as well) seldom act in the absence of organized public demand. Uncoordinated individual discontent, however widespread, has little effect on politicians and bureaucrats. Governments did not concede the political legitimacy of environmental quality concerns until citizen organizations with political muscle and sophistication emerged during the 1960s and grew in numbers and strength during the 1970s and 1980s (Hays, 1987; McCormick, 1989). In the United States, environmentalism gained force through the new right of citizen review of environmental impact statements provided by the National Environmental Policy Act, facilitated by the willingness of the courts to hear class-action suits and by the rise of public interest law firms (Caldwell, 1982).

Public environmental awareness, moving to apprehension and leading to organized pressure on public officials, occurred in nearly every developed country during the 1960s and 1970s, especially in Western Europe and North America (Inglehart, 1990; Milbrath, 1984). During these decades, many volunteer, nongovernmental environmental organizations became international or established international networks. They were present in large numbers at the Stockholm Conference, and an NGO Environmental

Forum was convened with the assistance of the Swedish government to provide an outlet for their complaints and agendas. This development helped conventionalize the involvement of environmental NGOs with official intergovernmental environmental policymaking. The relationship has continued and grown in association with the United Nations Environment Programme (UNEP), initiated in Stockholm (Caldwell, 1991; McCormick, 1989). The Environmental Liaison Centre at UNEP headquarters in Nairobi, Kenya, now interacts with more than 6000 environmental NGOs, representative of nearly every country on earth.

All of the foregoing developments have effected changes in both public opinion and national and international law. As previously noted, advances in science and technology have influenced environmentalism in varied ways and have extended the applicability of older statutes and conventions. For example, the technology of science has progressively led to refinements in chemical analysis. Pollutants (especially toxins) in air, water, and food can often be measured in parts per trillion, thus bringing into effect existing bans and restrictions regarding the presence of contaminants, in many cases contaminants whose presence had not previously been ascertainable. New laws and regulations have been adopted regarding the manufacture and distribution of new products. Concern for health has been a major impetus in environmentalism worldwide.[1]

Globalizing Popular Concern

During the years between 1965 and 1975 there was an upsurge of public awareness of the unity, finiteness, and interactivity of the planetary environment (Hays, 1987; Nicholson, 1970). The first landing on the moon on 20 July 1969 and pictures of the Earth from outer space brought to many people a realization that their environment had many of the characteristics of a closed system. "Spaceship Earth" became a metaphor, and "Only One Earth" was the motto of the 1972 United Nations Conference on the Human Environment. No specific action was indicated by this symbolism—the lives and prospects of people were not directly affected. However, by the mid-1980s, a series of highly publicized environmental disasters and revelation of environmental threats of global proportion brought home to people that nearly every environmental problem had personal and possibly international dimensions.

In recent years, several issues in particular have aroused concern that is leading to popular demand for both local and international action, including (a) transboundary transport of hazardous materials, chiefly wastes; (b) disintegration of the stratospheric ozone layer; and (c) global climate change, notably global warming. In each case, scientific findings, publicized in the news media, preceded public awareness and apprehension and eventually led to popular demands for governmental and international action. These were clearly not the types of issues that initiated U.S. environmentalism. Its first concerns were typically local or geographically restricted. In the United States, the legislative agenda during the 1960s and 1970s was characterized by measures to curb air and water pollution, to preserve wilderness areas and scenic rivers, and to protect specific endangered species.

Some commentators saw environmentalism as a largely American and transitory concern (Etzioni, 1970). However, during the decades since 1970, U.S. environmentalists have increasingly recognized the global dimensions of environmental problems and have identified their concerns with environmental movements around the world. Nearly every major U.S. environmental organization has an international component, and some (such as Friends of the Earth and the Sierra Club) have become international in membership. International organizations (e.g., Greenpeace) have become active in U.S. domestic environmental affairs, and growing numbers of U.S. citizens have joined it and other international coalitions (e.g., the Rainforest Action Network; McCormick, 1989). These developments reflect growing awareness that where formerly "foreign and domestic policies could be shaped in isolation, now they must be merged" (Mathews, 1989).

U.S. NGO examples, initiatives by the IUCN World Conservation strategy, and national appeals by the WWF have stimulated the formation and growth of environmental NGOs in developing countries. Many U.S. NGOs, such as the Nature Conservancy, recognize the need for counterpart organizations in countries where they work. The gap between countries over environment or development priorities has narrowed as the concept of environmentally sustainable development has gained adherents.

Informed citizen groups, almost invariably ahead of the public policy makers, have already formed links with action groups in developing countries. Thus, when incidents and developments abroad are introduced into discussions of U.S. environmentalism, it is not a departure from the focus on the United States. It is recognition that U.S. environmentalists are influenced by and concerned with world trends because their roles cannot be fairly described by attention only to their domestic agendas. Previously noted events such as the escape of toxic gas at Bhopal, India (1984), the diffusion of radioactive material from an accident at an atomic nuclear reactor at Chernobyl, U.S.S.R. (1986), a massive chemical spill in the River Rhine from the Sandoz chemical plant in Switzerland (1986), and the rupture of the oil tanker Exxon Valdez (1989) were covered at length and in detail in the U.S. environmental press.[2] Similar accidents in the United States have made it clear that the vulnerabilities are ultimately global, but these accidents are not as crucial for the global future as ozone depletion, global warming, and rainforest destruction.

Transboundary Transport of Harmful Materials

Expansion of world trade and transportation has carried with it an international flow of products, some of which have uncertain effects on health and environment. A phenomenal expansion of industrial chemistry, especially in relation to agriculture and medicine, has led to a large volume of regulatory protective legislation in the developed countries. In the United States, the Federal Environmental Pesticide Act of 1972 and the Toxic Substances Control Act of 1976 provided for government control over testing, manufacture, and use of chemicals believed to be dangerous. Similar measures have been enacted in other developed countries, but in the Third World protective legislation has as yet been minimal. In many of these countries, scientists and physicians have led efforts to restrict or at least identify hazardous materials imported from the industrialized world, and they have links with scientists and environmental organizations in the United States. Nevertheless, many less developed countries are threatened by dumping of hazardous wastes from industrialized countries and only recently have sought to reject this abuse. There is, moreover, the ethically dubious practice of exporting to less developed countries pesticides and pharmaceuticals banned in the countries of their origin (Caldwell, 1988).

The transboundary relationships needed to combat the spread of toxic chemicals have had a globalizing influence on the U.S. environmental movement. Faced with an intrusion of products whose safety has not been assessed and of substances banned for sale or use in their country of manufacture, Third World countries turned to the United

States and the United Nations for relief. In many Third World countries, governments lacked the competence and sometimes the integrity to enforce national protective measures. Thus, on 17 December 1981 the General Assembly of the United Nations, by a vote of 146 to 1, adopted Resolution 37/137, Protection Against Products Harmful to Health and Environment. The United States cast the only negative vote, to the chagrin of U.S. environmentalists.

Earlier, in 1980, hearings on export of hazardous substances had been held in the U.S. House of Representatives. Environmental, health, and consumer organizations advocated control measures, and, shortly before leaving office, on 15 January 1981, President Carter signed Executive Order 12264, which would have placed restrictions on the export of hazardous substances by U.S. companies. Soon after taking office, President Reagan withdrew Carter's order, with the argument that control of imports in other countries was their business and that the United States should not impose its values and standards on them (Smith, 1982).

In Europe, circumstances were different. Hazardous materials moved both ways across national borders. Efforts by the Organization for Economic Cooperation and Development (OECD) to develop transboundary controls were opposed by the U.S. (Reagan) delegate, reflecting the commercial interests of the U.S. chemical industry. Lobbying in Paris by the official U.S. representative against international protective measures was deplored by environmental groups in the United States and was resented by Europeans (Fouere, 1983).

Transhipments of solid and toxic wastes from industrial countries to poor countries of the Third World are often made without knowledge or consent of governments but increasingly are being monitored and exposed by environmental NGOs. For example, on 8 September 1987 the Greenpeace Toxics Campaign issued an International Action Alert protesting the shipment of toxic-laden incinerator ash from Philadelphia, Pennsylvania, to the remote province of Bocas del Oro in Panama. In Europe, attempts to dump hazardous wastes in Africa were frustrated by African governments that compelled the shipper to return the toxic wastes to their point of origin. In a similar vein, in 1987 a barge laden with 3168 tons of garbage from Long Island, New York, was forced to return to its home port after a 6000-mile journey through the Caribbean in a vain effort to find a country that, for a price, would accept the noisome refuse (*Facts on File,* 1987).

Such events, and many more, led, after protracted negotiations, to the adoption on 22 March 1989 of the Convention on the Control of Transboundary Movements of Hazardous Wastes and Their Disposal by representatives of 35 nations and the Commission of the European Community. The United States, Great Britain, the Federal Republic of Germany, and the Soviet Union did not sign but indicated their intention to do so, contingent on further study of the terms of the agreement. The waste management issue illustrates the progression of what was at first (and still is) a local issue becoming a national and then an international and global issue. The neglected problem of waste and contamination in worldwide, expanding industrial society has been a major cause of the globalizing of environmentalism and the conversion to environmentalism of hitherto unaware or indifferent people and governments.

Threats to Stratospheric Ozone

The stratospheric ozone issue was influenced more by public confidence in scientific findings than by direct experience. The issue arose with relative suddenness with discovery in the 1970s of a thinning of the ozone layer in the stratosphere. The ozone layer shields the earth from excessive ultraviolet radiation from the sun. Discovery of a hole in the ozone layer, first detected by British scientists over Antarctica, was soon paralleled by evidence of thinning over northern latitudes. The effect of increased ultraviolet radiation on human health and the biosphere was scientifically predictable with a high level of probability and credibility. Estimates of effects on food supply and on diseases of the skin and eyes were sufficiently alarming that governments responded with unprecedented alacrity.

An international treaty to restrict the emission of ozone-thinning agents, notably chlorofluorocarbons (CFCs), was signed by representatives of 20 nations meeting in Vienna, Austria, on 22 March 1985. Negotiations on more rapid phaseout of CFCs continued, and on 16 September 1987, 28 nations agreed to the Montreal Protocol on Substances That Deplete the Ozone Layer. Although by early 1989, 36 nations had signed the Montreal Protocol, dissatisfaction with the agreed-on timetable for phaseouts led to international meetings in March 1989 in London, England, and in May in Helsinki, Finland. Eighty-one nations were represented at the Finland meeting, which, on 2 May 1989, issued the Helsinki Declaration on the Protection of the Ozone Layer, calling on all states to join the previous ozone agreements and to phase out all production and consumption of CFCs as soon as possible, but not later than the year 2000 (Whitney, 1989; *Facts on File,* 1989). The United States and the European Community countries had already legislated to end the use of CFCs. Thus, here was a highly visible and decisive example of the globalization of environmental policy in which U.S. environmentalists had taken an early lead. The concentration of the manufacture and release of CFCs in a few industrialized countries made this issue easier to negotiate in an international forum than the more widespread and less clearly established issue of global climate change.

Global Climate Change

An issue of truly global proportions is the changing composition of the Earth's atmosphere, caused primarily by increasingly heavy emissions of carbon dioxide from the combustion of fossil fuels (oil, coal, and natural gas), methane (chiefly from agriculture), and CFCs. These emissions are residues of a rapidly populating and industrializing world. The effect of these so-called greenhouse gases in the atmosphere is to block the escape of infrared radiation (heat) from earth. If heat received from the sun is trapped on Earth, it will raise air temperatures and sea levels, melt the polar ice sheets, induce regional changes in rainfall, and vary the weather and the seasons. Increase in greenhouse gases in the atmosphere is exacerbated by massive burning of tropical rain forests for conversion to agriculture, with the consequential drop in emission of oxygen from a devegetated environment.

By the end of the 1980s, global warming and climate change had become the world's most publicized environmental issue. Here again, accumulating scientific evidence permeated the public consciousness and resulted in an unpredicted global political response. Scientists had been aware of the climate change issue for more than a decade before the issue "exploded" in the legislative chambers of governments. Scientific studies published in the early 1970s identified the effects of human activities on the climate, and in February 1979 the World Climate Conference meeting in Geneva, sponsored by the World Meteorological Organization, adopted a program of international cooperative research to more precisely identify the elements of the problem. In 1978 the Climate Research Board of the U.S. National Research Council released a report, *International perspectives on the study of climate and society: Report of the International Workshop on Climate Issues.* Other studies and reports followed in the United States and other countries, and in 1981 the issue came to the fore in the U.S. Congress, setting off a series of committee hearings throughout the rest of the decade.[3] Here is a paradox that has yet to be fully explored. Although there is general agreement that local issues (e.g., toxic waste dumps or solid waste incinerators) are principal foci of U.S. environmentalism, the global climate change issue has probably received more high-level governmental attention than any other single environmental concern. It is too early to say, but the global climate change–ozone issue may prove to be the greatest single stimulus for the globalization of environmentalism.

Other Global Issues and the U.S. Response

Other issues now perceived to be of global significance have aroused concern and action among U.S. environmentalists. Among these are the destruction of tropical rain forests (diminishing the diversity of plant and animal species and the production of atmospheric oxygen), desertification over growing areas (notably in Africa), acid rain, and threat to survival of the world's wildlife (including marine life, especially whales and dolphins). U.S. NGOs have mounted campaigns against tuna fisheries and Japanese trawlers whose nets have caused the deaths of countless numbers of dolphins. Networking with environmental NGOs in other countries, Americans have kept pressure on the International Whaling Commission for phaseout of commercial whaling. They have forced a reluctant federal government to ban drilling for oil in sensitive coastal areas and have thus far successfully blocked efforts to weaken endangered species protective legislation.

U.S. environmental organizations have also been in the forefront of *debt-for-nature swaps* (Cody, 1988). One of them, Conservation International, privately purchased some of Bolivia's foreign debt in return for promises of environmental protection for endangered natural areas. U.S. NGOs were not alone in pressuring the U.S. Congress and the World Bank for changes in policy for development loans that encouraged environmental destruction. British groups were also active (Goldsmith, 1987). However, the ability of Americans to influence the World Bank's leading source of finance (the U.S. government) enabled them to play a leading role in reorienting the World Bank and its development bank partners.

Guided by economists with strong inclinations toward monetary strategies, the banks had encouraged Third World nations to grow crops for sale abroad and to exploit natural resources to pay for international loans. The loans were often put to use in development projects that proved to be environmentally unsustainable. Many less developed countries were burdened by heavy foreign debt, which they sought to pay off by further exploitation of their natural resources and agriculture. Third World countries today have widely adopted the sustainable development concept as advocated by the U.N. Commission on Environment and Development but have linked their cooperation on environmental protection to relief from international indebtedness (Cody, 1988; McCormick, 1989).

Although publicity, political pressure, and negotiations have been the principal strategies of NGOs, an ultimate goal of U.S. environmentalism has been to institutionalize its objectives in law. The growth of national environmental law in the United States, Canada, and Western Europe, and, in some respects, in Japan during the 1970s and 1980s has been phenomenal. In at least thirty-seven countries, statutory law has been to some extent reinforced by constitutional provisions.[4] International law has undergone a somewhat comparable expansion in the form of new conventions and other agreements and in the establishment of implementing agencies and programs that give international commitments greater force. Americans as private individuals and through the American Society for International Law and through World Peace Through Law have been leaders in the development of international law, including the U.N. Law of the Sea Treaty. They have often been more international in orientation than their government, especially during the 1980s when official U.S. policy on environmental affairs was (with exception of the ozone issue) largely out-of-step with that of the rest of the world.

Realization of the Stockholm legacy occurred progressively but unevenly over an interval of 17 years. Nevertheless, by the July 1989 Paris summit meeting of the political heads of the seven major industrial democracies, the environment had become a top item on the world's political agenda. The threats of global climate change and of disintegration of the stratospheric ozone layer were the primary developments commanding attention at the highest international political levels. However, cumulative experience with international conferences, declarations, agreements, investigations, and cooperative programs since Stockholm made the Paris Communique of 7 July 1989 more of an overdue catch-up by chiefs of state than an excursion into the unknown (Markham, 1989). How high-level commitments were to be translated into action in the 1990s could not be foreseen, but global environmentalism was never more pervasive and Americans never more engaged in international environmental affairs.

Nevertheless, the commitment of the Bush presidency to the environment was ambiguous. Although candidate Bush made a major point of his environmental concern, the policies of President Bush often have seemed contradictory. The appointment of William Reilly as administrator of the Environmental Protection Agency was warmly applauded by environmental organizations. However, White House Chief of Staff Sununu, Office of Management and Budget Director Darman, and Interior Secretary Lujan are frequently described in the news media as unsympathetic to what they regard as excessive environmental concerns. Who speaks for the President? As of 1990 there was considerable doubt in the United States and abroad regarding the depth of the environmental commitment of George Bush, but even the skeptics consider him a great improvement over his predecessor.

Implementing Global Environmentalism: Political and Institutional Challenges

A phenomenon of the 1970s and 1980s, as previously noted, has been the burgeoning of nongovernmental environmental organizations with international agendas (McCormick, 1989). There are many nongovernmental organizations around the world with many causes but it is unlikely that any exceed the environmental organizations in the United States in membership, in diversity, and in their commitment to international causes. NGOs from many nations have had an unofficial but organized presence at every United Nations environment-related conference since Stockholm, including those on population, food, science and technology for development, and new and renewable sources of energy. Links between these unofficial convenings and the actions of official national representatives may not be visible during these occasions. For some countries, notably the United States, there have been contradictions between positions taken by official delegations and the policies advocated by a majority of the NGOs present. This was frequently the situation during the Reagan years, and even some official representatives of the United States found themselves unable to support the President's positions.

On home fronts, environmental NGOs may press their governments for or against compliance with conference resolutions. Coalitions of NGOs have been formed to promote global environmental policies, and international nongovernmental conferences have been held in Europe and North America in which the agenda has been global (e.g., Globescope in the United States, sponsored by the Global Tomorrow Coalition). In Europe, especially in countries with proportional representation in their elective legislatures, transnational voting for candidates to the Parliament of Europe facilitates transboundary collaboration among green parties and other environmentally concerned groups.

The principle of proportional representation, in effect in many European democracies, enables small parties to elect representatives to local and national legislative bodies. The result has been multi-party politics in which relatively minor parties may hold a balance of power when none of the major parties have sufficient votes to form a government. Green or ecological parties have emerged and have gained considerable influence over the traditional parties (Capra and Spretnak, 1986). The U.S. political system has been uncongenial to third- or fourth-party movements. U.S. environmentalists have fought or favored individual politicians on their environmental records, regardless of party affiliation. Should there be a break-up of the two-party system at some future time, environmentalism might become a more coherent force in U.S. politics.

Is global environmentalism a response to a climacteric in world affairs? If so, what might this signify for the future of U.S. politics? In human life, a *climacteric* is the point at which physical powers, having reached their climax, begin to decline. The human experience in populating the earth and in exploiting its resources may be analogous. During modern civilization's exuberant youth, roughly since 1492, it was possible for man to turn nature to his own purposes without regard to ecological or human consequences. Now, after half a millennium of unrestrained populating and developing the Earth, it seems that modern society has gone about as far as it can go with the policies and methods that it has heretofore used. The carrying capacity of the Earth in relation to humans may have been exceeded already (Catton, 1980). More care and caution will henceforth be required to sustain the human economy without progressive decline in the ability of the environment to maintain life on Earth.

Environmentalism in North America and throughout the industrially advanced world generally may be understood as a part of a more general shift in perceptions and values. During the last half of the twentieth century, parallel societal changes seem to have occurred spontaneously in North America, Western Europe, and other developed nations (Inglehart, 1990; Milbrath, 1984). Manners and morals, attitudes toward freedom of speech, dress, human rights, and lifestyles have followed similar courses, and growing concern for the environment has been a major dimension of the change. Globalizing of U.S. environmentalism is merely a particular case of the globalizing of environmentalism everywhere. It is a manifestation of a worldwide trend.

Global Environmentalism and Sustainable Development

In the course of history, global environmentalism is a sequel to the globalization of international trade and commerce, a process that began in the seventeenth century with British and Dutch enterprises in America, India, and the Far East. Colonialism was a globalizing process continued in our own time through reverse colonialism, wherein an overpopulated Third World overflows into the developed countries. To stem this flow, to improve the prospects for life in the Third World, and to save the biosphere, global environmentalism today has adopted the concept and goal of sustainable development (Redclift, 1987; Tolba, 1987). The report of the World Commission on Environment and Development (Bruntland Commission), entitled *Our common future* (1987), has become the closest approach to a scripture for global environmentalism. The report declares for sustainable development and undertakes to reconcile the environmental and developmental interests of developed and developing countries. The United States government (unlike those of many other countries) has chosen to ignore the report, which, however, is used widely in environment and development courses in U.S. colleges and universities.

It is anticipated that the Bruntland report will provide a conceptual basis for the projected United Nations Conference on Environment and Development to be held in 1992 in Brazil. By then the draft of an international convention on the atmosphere is expected to be ready for review. If by that time the Convention on the Law of the Sea has obtained the necessary ratification (for this, U.S. adherence is probably essential), human use of the two elements that encircle the globe will be brought under comprehensive international law. Dozens of treaties and other international agreements cover many of the uses of the land, its biota, and its inland waters. Global environmentalism will no longer be regarded as exceptional because, at least in context and awareness, all environmentalism must henceforth be global. Nor will it be possible for the government of the United States, as it did during most of the 1980s, to remain aloof and often negative toward environmental issues that a large part of the world regards as requiring global cooperation.

Notes

1. It is paradoxical that a concept as broadly and diversely holistic as *environment* has come to be treated as an "ism"—a sectoral or particularistic way of relating to practical affairs. Environmentalism is not a self-designating term used by the environmentally concerned. It has been applied often, with a pejorative undertone, among those interests in modern society in which more than a minimal concern for the environment is regarded as atavistic, impractical, utopian, and unprogressive. Thus, the broader ecological perspective is treated as if it were narrow, and a narrow economistic orientation, where dominant, tends to be regarded as practical and realistic.

2. For summary accounts of disasters receiving global news coverage, see *Keesing's contemporary archives: Record of world events*: Seveso (June 1979:29688; October 1983:32475; March 1986:34270), Bhopal (March 1985:33467–33468), and Chernobyl (June 1986: 34460–34462). See *Facts on File* for Rhine River (7 November 1986:845G1; 14 November 1986:858E1; 31 December 1987:89B2), and for Exxon Valdez (March 1989:36541, 36606). See also *Challenges for international environmental law: Seveso, Bhopal, Chernobyl, the Rhine and beyond,* edited by V. P. Nanda and B. Bailey (Washington DC: World Peace Through Law Center, 1987).

3. There is an extensive literature on the greenhouse effect and global climate change. See especially *The potential effects of global climate change on the United States,* by the Environmental Protection Agency (October 1988), a draft report to Congress (Washington, DC: Author); the special issue of *Science* on "Issues in atmospheric sciences" (10 February 1989); the special issue of *Scientific American* on "Managing Planet Earth" (September 1989); and *Global warming: Are we entering the greenhouse century?* by S. H. Schneider (San Francisco: Sierra Club Books, 1989). For Congressional hearings and reports, see, for example, the following:

U.S. Congress, House of Representatives. *Carbon dioxide and climate: The greenhouse effect.* Hearing before the Subcommittee on Natural Resources, Agriculture and Environment and the Subcommittee on Investigations and Oversight of the Committee on

Science and Technology, 97th Congress, 1st Session, 31 July 1981. Note testimony of Stephen Schneider, "Carbon dioxide and climate: Research on potential environmental and societal impacts," pp. 39–59. Note also hearings before these Subcommittees, 98th Congress, 2nd Session, 28 February 1984.

U.S. Congress, Senate. *Global warming.* Hearing before the Subcommittee on Environment and Public Works, 99th Congress, 1st Session, 10 December 1985.

U.S. Congress, Senate. *Ozone depletion, the greenhouse effect, and climate change.* Hearings before the Subcommittee on Environmental Pollution of the Committee on Environment and Public Works, 99th Congress, 2nd Session, 10–11 June 1986.

U.S. Congress, Senate. *Ozone depletion, the greenhouse effect, and climate change.* Joint hearing before the Subcommittee on Environmental Protection and Hazardous Wastes and Toxic Substances of the Committee on Environment and Public Works. 10th Congress, 1st Session, 28 January 1987.

U.S. Congressional Research Service, Library of Congress. 1984. *Carbon dioxide, the greenhouse effect, and climate: A primer.* Report, transmitted to the Committee on Science and Technology, U.S. House of Representatives, 98th Congress, 2nd Session, October 1984. Washington, DC: U.S. Government Printing Office.

See also *Society and Natural Resources,* vol. 4, no. 4, a special issue entitled "Global climate change: A social science perspective," ed. Martin F. Price, published in December 1991.

4. Information provided by the International Environmental Law Centre, Bonn, Germany. A slightly different list is provided by Edith Brown Weiss in *Fairness to future generations* (Dobbs Ferry, NY: Transnational, pp. 107–108). The total of combined lists is 45 countries in which some reference to the environment, natural or cultural, is made. In no more than a third of these, however, is the provision likely to have practical significance at present.

References

Caldwell, L. K. 1982. *Science and the National Environmental Policy Act: Redirecting policy through procedural reform.* University: University of Alabama Press.

———. 1984. The world environment: Reversing U.S. policy commitments. In *Environmental policy in the 1980's: Reagan's new agenda,* ed. N. Vig and M. Kraft, pp. 319–338. Washington, DC: Congressional Quarterly Press.

———. 1988. International aspects of biotechnology: Guest editorial and review. *MIRCEN Journal of Applied Molecular Biology and Biotechnology,* 4:245–258.

———. 1990. *International environmental policy: Emergence and dimensions,* 2nd ed. Durham, NC: Duke University Press.

Capra, F., and C. Spretnak. 1986. *Green politics.* Sante Fe, NM: Bear and Company.

Carson, R. 1962. *Silent spring.* Boston: Houghton Mifflin.

Catton, W. R. 1980. *Overshoot, the ecological basis for revolutionary change.* Urbana: University of Illinois Press.

Cody, B. 1988. *Debt-for-nature swaps in developing countries.* Washington, DC: Congressional Research Service.

Dinwoode, D. H. 1972. The politics of international pollution control: The Trail Smelter case. *International Journal,* 27:219–235.

Etzioni, A. 1970, 22 May. The wrong top priority. *Science,* 168:921 (editorial).

Facts on File. 1987, 17 July. Vol. 47, p. 518.

———. 1989, 12 May. Vol. 49, pp. 334–335.

Fouere, E. 1983. Clashing over the environment: U.S., Europe follow increasingly divergent paths. *Europe,* 237:12–15.

Goldsmith, E. 1987. Open letter to Mr. Conable, president of the World Bank. *The Ecologist,* 17:58–61.

Hays, S. P. 1987. *Beauty, health, and permanence: Environmental politics in the United States, 1955–1985.* Cambridge, MA: Cambridge University Press.

Inglehart, R. 1990. *Culture shift in advanced industrial society.* Princeton, NJ: Princeton University Press.

Livingston, D. 1968. Pollution control, an international perspective. *Scientist and Citizens,* 10:173–182.

Markham, J. M. 1989, 12 April. Greening of European politicians spreads as peril to ecology grows. *New York Times,* p. 10.

———. 1989, 17 July. Paris group urges "decisive" action for environment. *New York Times,* pp. 1, 4–6.

Mathews, J. T. 1989, 11 June. Forging a policy to address global warming. *New York Times,* p. 29.

McCormick, J. 1989. *Reclaiming paradise: The global environmental movement.* Bloomington, IN: Indiana University Press.

Milbrath, L. 1984. *Environmentalists: Vanguard for a new society.* Albany: State of New York University Press.

Nicholson, E. M. 1970. *The environmental revolution.* New York: McGraw-Hill.

Nisbet, R. 1982. *Prejudices: A philosophical dictionary.* Cambridge, MA: Harvard University Press.

Petulla, J. M. 1980. *American environmentalism: Values, tactics, priorities.* College Station: Texas A&M Press.

Ramphal, S. S. 1987. The environment and sustainable development. *Journal of the Royal Society of Arts,* 135:879–909.

Redclift, M. 1987. *Sustainable development: Exploring the contradictions.* London: Methuen.

Rosen, G. 1964. Human health, community life, and the rediscovery of the environment. *American Journal of Public Health and the Nation's Health,* 44(Part 2):1–6.

Smith, R. J. 1982. Hazardous products may be exported. *Science,* 216:1301.

Tolba, M. K. 1987. *Sustainable development: Constraints and opportunities.* London: Butterworth.

U.S. National Research Council. 1978. *International perspectives on the study of climate and society* (Report of the International Workshop on Climate Issues). Washington, DC: Author.

Whitney, C. R. 1989, 3 March. 12 European nations to ban chemicals that harm ozone. *New York Times,* pp. 1, 4.

World Commission on Environment and Development. 1987. *Our common future.* New York: Oxford University Press.

Twenty Years of Change in the Environmental Movement: An Insider's View

MICHAEL McCLOSKEY

Sierra Club
730 Polk St.
San Francisco, CA 94109
USA

Abstract *In examining changes in the U.S. environmental movement since it emerged 20 years ago, I conclude that the movement has now split into three camps: a radical segment, a mainstream segment, and a segment anxious to seek accommodations with industry. These camps are distinguished by their goals, their attitudes toward government and industry, and the means they use. Other changes have also become clear: weakness in mobilizing for issues other than nature protection, a decline in campaigning capabilities, a lack of articulated vision, and difficulty in getting environmental statutes implemented. Growth in membership and range of activities, as well as public support, have masked these problems. Mainstream groups are confronted with criticism that they can no longer produce real improvements in environmental quality through governmental action. A way to ease the tensions between the camps is suggested, that is, by having the mainstream groups borrow suggestions from the other two camps and focus energies on green consumerism as a way to shift emphasis to a more productive arena, that of directly influencing corporate behavior.*

Keywords Environmental ideology, environmental movement, environmental strategies, environmental tactics, management of environmental organizations, splits in environmental movement.

The environmental movement in the United States has changed substantially over the past 20 years, but the change has been so gradual that few recognize how much change has occurred and what profound questions are posed by this change. This essay examines the extent of these changes and, in the process, assesses both the overall strength of the movement and the strains that afflict it. Specifically, I discuss changing ideological cleavages within the movement, changes in the key issues addressed by the movement, and shifts in movement strategies and tactics over the last 20 years. I then describe how the major national organizations such as the Sierra Club changed over the two decades and conclude by discussing what I see as major challenges facing the environmental movement.

Ideology

When the contemporary U.S. environmental movement came together in 1970, it was clear that camps existed within it (and, early on, those camps did communicate with one another). The dominant camp comprised the pragmatic reformers, who are still most prominent today. They believed that progress toward environmental protection could be achieved best through government action, and they focused their efforts on influencing public policy in incremental steps, forging pragmatic alliances issue by issue with those with whom they could agree on a given issue. They focused on government because it afforded the greatest leverage by changing the rules of the game for everyone, and they felt that timely progress could best be made in that way. They did not believe that entire political and economic systems needed to be changed and were confident that environmental protection could be achieved within the framework of existing institutions of governance.

The pragmatic reformers were challenged by a lifestyle camp. Those in this camp had little confidence in government and looked mostly to what individuals could do in their own lives to protect the environment. By living frugally and simply, individuals could change patterns of consumption. In that fashion, they hoped to reduce pollution and the drain on resources; they also hoped to pioneer in the use of appropriate technologies. Those in this camp also believed that decisions should be made in a highly participative and even consensual manner.

The split into these two camps was somewhat obscured by the excitement of the Earth Day period, and indeed many environmentalists shared some of the two contrasting ideologies. Some have, somewhat invidiously, contrasted these camps in terms of power and participative strategies (Morrison, Hornback, and Warner, 1972). The potential for friction was there but was avoided for the most part by the quick fading of the lifestyle camp (e.g., the rapid decline of "voluntary simplicity"). Many in this camp moved into rural communes, and others went on to other things as the youth rebellion of the 1960s died. A fragment of this camp persisted in the protests at nuclear power sites mounted by the various alliances, such as the Clamshell Alliance.

From about 1972 on, the environmental movement was remarkably free of stress over ideology. However, this changed again by the mid-1980s when a new radical wing emerged in the environmental movement. This new radical wing was something more than a revival of the old lifestyle wing. It had many more strands and was characterized by a strong critique of the conventional methods of the pragmatists (Borrelli, 1988). To some extent, this new radicalism embodied a reaction against the anti-environmental radicalism of the Reagan administration. It reflected a determination to go as far as possible in the opposite direction.

Its emergence also coincided with the decline of anti-establishment groups within the pragmatic camp—groups such as Friends of the Earth (FOE), the Environmental Policy Institute (EPI), and Environmental Action. All faced increasing problems in raising money and continuing operations. FOE and EPI merged, and Environmental Action merged with the Environmental Task Force. One can infer that those who wanted strong words found them more satisfying when they came from groups that had given up on the government, and consequently they joined and donated to more radical groups instead.

Some of the new radicals were radical in their demands, and others were radical in the means they used. The deep ecologists, the bioregionalists, and the residue of the lifestyle camp wanted sweeping changes in society and living patterns but were largely apolitical. Their demands were radical, but their means did not shock people. Also radical in their demands (but less so) and not shocking in their means were those organizing green party units, followers of Barry Commoner, social ecologists, and local radical activists such as groups mobilized around toxics issues and neighborhood "not in my back yard" (NIMBY) groups (Bookchin, 1990; Borrelli, 1988; Freudenberg, 1984).

Some groups had demands that were not at all radical, but they used direct-action techniques of protest that seemed radical in their confrontational style. Groups such as Greenpeace prospered in using these techniques, and smaller groups such as the Rainforest Action Network did too. In a sense, this was really a continuation of the approach pioneered by the nuclear protesters, and indeed, Greenpeace began protesting nuclear explosions.

Finally, groups such as Earth First! and the Sea Shepherd Society actually used radical means, resorting to sabotage and other illegal techniques, although their demands for changes in public policy were not always radical (Borrelli, 1988). It is interesting, too, that these groups were still concerned with public policy and often addressed rather conventional issues of preserving forests, whales, and other wildlife.

Most of the radicals took pains to distinguish themselves from the mainstream, pragmatic groups. They broke the long-standing rule of not speaking ill of their brethren; they did not see themselves as brethren but as stern critics pointing out the error of the ways of the mainstream groups. These latter groups were attacked for being wrongheaded in placing their faith in a government that had betrayed them. Many radicals wanted to attack the basic system of industrialism and consumerism, and the mainstream groups did not. The mainstream groups were attacked for not getting the job done—for being complacent, co-opted, bureaucratic, distant, arrogant, interested only in professional "perks" and money, and for being too conservative. Needless to say, these attacks, which were largely not reciprocated, ended any sense of comity or unity in a common cause. They marked a new cleavage that seems now as great as the old split between the Muir and Pinchot traditions of the conservation movement (Hays, 1959).

Less pronounced, but nonetheless clear, was another split on the other end of the spectrum. Fred Krupp, the executive director of the Environmental Defense Fund, had announced in the mid-1980s that the era of confrontation was over. He predicted that a new era was beginning in which industry and environmentalists would work together harmoniously; this would be a time of accommodation. These accommodators would look less to the heavy-handed governmental regulation favored by the reformers and more to market-like mechanisms to achieve their ends. This rationale gave more establishment-oriented groups such as the Conservation Foundation, Resources for the Future, the World Wildlife Fund, the Nature Conservancy, and the World Resources Institute a nicely articulated ideological niche.

Although they rarely gave voice to their criticism of the pragmatic reform groups, the accommodators established their distance and rarely collaborated with these groups. Some organizations, such as Resources for the Future, provided a home for economists who critiqued the conventional dogmas of mainstream groups in the pollution control field. These critiques are carried even further by right-wing theorists in think tanks such as the Heritage Foundation and the Cato Institute, which want to privatize the public domain and regulate pollution through tort law.

Thus, at the end of the 1980s the mainstream groups found themselves under attack by critics on either side—critics who had no interest in collaboration. It was also clear that groups in the other camps were attracting a following and growing in size. Greenpeace claimed over one million followers in the United States, as did the World Wildlife Fund; the Nature Conservancy passed the half million member mark, and they all became major players. With its new positioning, the Environmental Defense Fund also was attracting significant funding. Mainstream groups such as the National Wildlife Federation and the Sierra Club were growing too, but the public was not saying no to those espousing divergent strategies.

Issues

Over the years, the issues of concern to environmentalists changed as well. If anything distinguished the post–1970 environmental movement from what preceded it, it was its holism and the breadth of issues it embraced. Yet, the broadening issues were having a hard time establishing staying power. Population growth and growth control (land use) issues helped usher in the environmental movement (along with pollution control), but they proved to be perenial stepchildren among the mainstream groups. At the end of 20 years, they were not top issues for any of these groups. Neither were the energy issues that hit so hard in the 1970s. When the global warming issue arrived at the end of the 1980s and energy issues became pertinent again, there was barely a corporal's guard of competence left in the movement on these once pivotal issues. The nuclear winter issue also came and went with great fanfare in the early 1980s.

Even the pollution issues, which were at the heart of the modern movement, struggled for a constituency. In the early 1970s, they were top–down issues. They were pushed by the national leadership as an exercise in intellectual commitment, but it was hard to build excitement in the ranks. The issues were too complex and enmeshed in jargon; some wondered whether pollution laws would ever get implemented when there were so many steps in the process where they could fail. Sometime in the late 1970s, these issues began to acquire a grassroots following. People in localities began to feel threatened with toxic wastes, carcinogens, and other hazards. Accordingly, the issue became more of a people's issue. Nevertheless, this activism expressed itself largely through local, ad hoc groups that focused on immediate health hazards. At the national level, pollution issues became boring. Too much complexity was summed up in the nationwide picture and the various pollution control laws, and the regulatory processes were a nightmare that no one could fathom.

The net result of this was that the best rallying-cry issues of the mainstream movement continued to be public lands and nature-protection issues. Groups such as the Sierra Club put tremendous resources in the 1980s into the pollution issues—defending and strengthening the Clean Air Act, expanding Superfund, getting the Clean Water Act reauthorized, and cleaning up defense production facilities. However, the issues that captured the hearts of its members were protecting wilderness, defending Alaska's wildlife, saving old-growth forests, and "keeping the money changers out of the temples." The polls showed that people cared about the pollution issues, but that was not the same as caring about the intricacies of the Clean Air Act. The nature-protection issues were imageable—in your mind you could see what you were saving and imagine how it would be saved. You could commit yourself to that kind of campaign. That was not true with most of the other broader issues.

When the mainstream groups put their heads together to develop the *Environmental agenda for the future* (Adams et al., 1985) and the *Blueprint for the environment* (Maize, 1988)—both vision statements for the future from the mid and late 1980s—they were shocked at how thin their collective expertise was on the pollution issues. In most cases, it came down to a few staff members at the Natural Resources Defense Council (NRDC) and sometimes the Environmental Defense Fund (EDF; Adams et al., 1985; Maize,

1988). These issues simply had not been institutionalized into the warp and woof of the movement, at least not in terms accessible to those in the national offices.

Strategies

These problems within the environmental movement were masked to some extent by the deteriorating relations with Presidents and their administrations. Despite initial skepticism, environmentalists came to accept progress under the Nixon and Ford administrations as normal, and they were elated by the strong commitment they perceived in the Carter administration (though this changed in the end). However, the strong hostility of the Reagan administration, which persisted for 8 years (although the hard edge was taken off at the end of the first term), stunned the movement. Not only did the federal government no longer propose new initiatives, it no longer even tried to maintain the programs of the past. Reaganites schemed to undo and dismember pollution and public lands programs. Normal diplomatic relations with administration figures virtually ceased. The movement lost any incentive to temper its criticisms because it knew that the administration was in the hands of ideological opponents who were not only unsympathetic but antagonistic. Adversarial relations became normal (Friends of the Earth, 1982).

However, the barrage of criticisms that the movement unleashed toward the administration had unexpected results. First, the movement came to expect less and less as normal, which lowered the threshold for acceptable performance. And second, the criticisms not only hit the Reagan appointees but the federal government itself. The anger and frustration over the federal government's abandonment of environmental programs spilled over to attach to the career bureaucracy and allied institutions. The career people in agencies such as the Environmental Protection Agency (EPA) were seen as not having kept the faith and as having been subverted by notions such as EPA is a risk-assessment agency instead of a pollution-reduction agency. Other natural resource agencies were seen as lacking the will to do anything right anymore and as being in a perpetual state of war with the public they served. None of the federal agencies was seen as caring about the views of the constituencies who had created them.

This very frustration also created the tension implicit in the performance gap. Public anger over the default of the federal government expressed itself in polls showing incredibly high levels of demand that something positive be done (Dunlap, 1987, 1989). There was a tremendous gap between what the public wanted and what it was getting. The public felt so strongly because the federal government had turned its back on their demands, injuring its status in the eyes of the public. Disillusionment with the federal government may have been exactly what the Reaganites were seeking to engender, but it posed a problem for a movement that had always placed so much emphasis on the federal role. Would its constituents continue to rally to calls for new federal programs?

In the meantime, a slow reversal had occurred with regard to the roles of state and local government. When the movement began, state and local government was seen as largely in the hands of those with little sympathy for environmental goals. However, by the end of the 1970s this had begun to change, with environmentally sympathetic administrations coming to power in many states and localities. Now the movement came to see these jurisdictions as places to innovate and set the pace and to outflank federal intransigence. Industry, in turn, came to hope the federal government would preempt the field with weak standards and prevent states from setting higher standards. This reversal in attitudes, however, reflected falling expectations with regard to what could be accomplished at the federal level.

It also highlighted the confusion over where to go to seek solutions. From the outset the movement had been caught in the conflict between its philosophy and what was happening in the world with respect to decentralization. Its Earth Day era philosophy favored decentralization and breaking institutions into smaller units, which were more understandable and could be more responsive. However, as the world economy became more interconnected and technology introduced greater complexity, the environmental movement found itself chasing after new sets of problems, such as global warming and damage to the ozone layer. Dealing with such problems called for more centralized planning and international treaties and for ever greater complexity in the scale of organization. The scale of the problems demanded commensurate solutions, but that led away from decentralization, at least in the short run, and took the movement even further from its grassroots and policies its supporters could relate to.

This problem, moreover, exemplified what was happening as time passed after Earth Day. Connections to the philosophy that gave birth to Earth Day were becoming more and more tenuous, and the broad visions of that time (1970) were seldom articulated anymore. Few could see what all the solutions would add up to or where the amelioristic solutions of the pragmatists were heading. Indeed, pragmatists found reasons for not projecting too far ahead for fear of alienating would-be allies on given issues and for fear of being wrong on the scientific basis of newly emerging problems. However, by the same token, this lack of utopian vision left the new recruits, who were mobilized by the anti-Watt wars and the seriousness of the new threats, unsure of what flag they were following. They knew that the condition of the environment should be improved, but they could not tell what kind of society could best provide that improvement.

Tactics

Over some 20 years, the environmental movement has steadily improved the toolbox of techniques it can use to influence public policy. Lobbying had been used since the beginning of the conservation movement in the 1890s, along with publicity campaigns (Fox, 1981). Litigation was added as another basic tool in the 1970s, with the movement having hundreds of cases pending at any one time. Late in the 1970s, as the movement realized that it needed to focus more attention on getting its friends re-elected and getting more such people into Congress, more groups organized political action committees (PACs). The League of Conservation Voters and Environmental Action had pioneered in endorsing candidates earlier in the 1970s, but the scope of the work exploded in the 1980s as the Sierra Club came to endorse as many as 200 candidates for public office at each election and spent millions in that decade for this purpose.

The conduct of campaigns also became more professionalized as computers and laser printers were used to mobilize constituencies. Computerized membership lists were broken down by Congressional districts; phone banks routinely prompted calls to Congressional offices; mail-grams to thousands of members could be generated by an electronic signal. Waves of constituents were brought in to lobby Congress, and flyers went out almost with every mail. If anything, the process became almost too routine. Congressional offices were deluged with mail and began to need to see more mail to feel moved. Moreover, the mail needed to hit after "their" mail had arrived, and the impact of mail wore off quickly. Members of these organizations dragged in for lobbying rounds often felt like cannon fodder.

So many last-ditch campaigns were run by the Sierra Club that campaigns lost their sense of being special. This was especially true with the spate of reauthorizations of pollution laws. None of them was final; there would always be another one. Lobbying campaigns lost their intensity and sense of drama as history in the making. Memories of major efforts faded quickly. More ominously, few organizations other than the Sierra Club were even trying to mount major campaigns anymore.[1] The campaign that involved a large budget, great specialization in terms of staff roles, mass publicity, and continuing constituency mobilization became rarer and rarer.

Faith began to wane that major lobbying efforts had much payoff anyway. Surely, the movement could still get statutes enacted, and it could win lawsuits over government refusal to implement them. But could we get them implemented properly in the final analysis? As in the adage, we could lead a horse to water, but could we force it to drink? Litigation and Congressional oversight could move a recalcitrant executive branch only so far; they could never force it to implement a law well.

By the end of the 1980s there was a sense that environmental regulatory programs were a shambles, with much of the promise of environmental statutes reduced to dead letters. Inconvenient environmental language was simply ignored or treated as Congress's opinion. EPA programs seemed to be all input and virtually no output. What was the point of great lobbying campaigns in Congress if so little came of the enactments in the end?

How the National Organizations Have Changed

The movement's narrowing focus and diminishing commitment to campaigning as the 1980s ended was obscured by the success in attracting a following that most environmental groups enjoyed over these two decades. Twice as many new groups were founded as went out of business. Memberships doubled or trebled; budgets grew sometimes by as much as tenfold (as in the case of the Sierra Club). A flurry of entrepreneurialism got many of them into enterprises of various sorts: product catalogs, publishing, running tours, issuing credit cards, and consulting. Running these organizations became complex and came to be a management challenge, and this was all the more true for those groups with participating members who were organized into chapters. The fund raising and membership recruitment work alone became very professionalized, as leaders began to work with list brokers, copywriters, foundation funders, and deferred-giving specialists.

Those running such organizations often found difficulty making time to set strategies for the environmental work to be done. It was left increasingly to issue specialists who knew one area but had little experience that would give them an overview. Fewer and fewer chief executive officers (CEOs) were hired from within environmental organizations or even out of the ranks of environmental experts. Most were hired from the outside because of their abilities in public relations, work with funders and donors, and management experience. They had not learned the environmental business on the firing line. They have no institutional memory on which to draw (and, with turnover, few are left with this memory in many organizations).

Such environmental CEOs were loath to question accepted ways of doing business and have largely been content to keep things going as they have been, at least with regard to strategies and issues. As managers, many have put emphasis on putting management procedures in place, on working through channels, and on staying within the budget. This has added to the rigidity of the organizations and has made it harder for them to innovate and to evaluate which approaches are most productive.

History shows that organizations typically enjoy greatest success when they manage to combine elements that are often uncomfortable with one another: those who have a passion for the mission and are driven by their own visions and those who can manage finances and personnel well. Environmental organizations with visionary leaders have attracted great followings and then fallen deeply into debt and bad times; others with stolid management have attracted little following because they generated no excitement (Fox, 1981).

At the moment, groups with little sense of vision are prospering simply because of the fund-raising talents of those working for them and because of the receptivity of the market. The same direct-mail consultants often have contracts with a number of organizations, and their copywriters' skills bring in membership for them all. But this success does not necessarily mean that the organizations are running successful environmental programs. The success speaks more for the skills of the consultants and the hunger of the public for hope. A crisis lurks behind this facade: Do these organizations know how to deliver what the public thinks it is getting—a better environment?

Many of the organizations are very good at certain things: the Sierra Club in mobilizing its large cadre of grassroots activists; the Wilderness Society at research and public relations; the Nature Conservancy at habitat surveys; the National Wildlife Federation in reaching middle-class and working-class people with an environmental message; the NRDC in mastering the complexities of federal programs; the Audubon Society in running sanctuaries; and the Sierra Club Legal Defense Fund in litigating. Nevertheless, these strengths do not necessarily add up to success for a movement, particularly as times change. None of these organizations, for instance, has shown any great expertise in lobbying administrative agencies, although NRDC knows how to talk to agency specialists.

As the movement became embattled with the Reagan administration, the mainstream groups began to collaborate more. The "Group of Ten" (now embracing more than a dozen groups) was formed so that the CEOs of the more active groups could meet quarterly to coordinate their strategies, particularly in dealing with a hostile administration. The Conservation Directors (the so-called "B Team") of these groups also met regularly to coordinate campaign planning. Various ad hoc coalitions and task forces met as needed, with some, such as the Clean Air Coalition, carrying on for years. Occasionally, competing task forces emerged where splits occurred over approaches.[2] Quite often, groups outside the movement, such as the American Lung Association, the Steelworkers Union, the American Civil Liberties Union, and the National Taxpayers Union, were brought in to broaden the base. Various umbrella groups, such as the Natural Resources Council of America, brought professional societies and trade associations into contact with activist groups.

By the end of the 1980s, the movement also found to its surprise that other movements were picking up the environmental theme and institutionalizing it within their organizational structures. The AFL-CIO organized the OSHA-Environmental Task Force to promote cooperation with respect to workplace and community exposure to environmental hazards. Consumer groups moved into the arena not only through Ralph Nader's considerable network, but also through the Consumer Federation of

America as it hired energy and environmental specialists. The American Association of Retired Persons (AARP) did likewise and promoted active energy conservation programs at the state level. The National Council of Churches organized its Eco-Justice Working Group to focus on the interface between environmental protection and social justice. Various scientific groups moved into this field, including the American Federation of Scientists and the Union of Concerned Scientists. Many U.S. businesses hired environmental specialists to advise them and reached out to set up dialogue and mediation projects whenever Washington began to stress environmental issues. In 1989, social investment companies banded together to launch a campaign to pressure firms to agree to pledge to abide by ten principles of good environmental practice (the Valdez principles). As the 1980s closed, even the long-troubled relations between the environmental movement and the Civil Rights movement began to show signs of changing for the better, as some blacks began to explore ways of finding more jobs for blacks in environmental work.

The spreading involvement of other groups in environmental issues did not prevent some competition between mainstream environmental organizations. Almost every group faced a logical competitor trying to occupy the same market niche and competing for visibility, leadership, membership, and funds. NRDC competes in this way with the Environmental Defense Fund, the Nature Conservancy with the Trust for Public Lands, the Wilderness Society with the Sierra Club and the National Parks and Conservation Association, Friends of the Earth with Environmental Action, Defenders of Wildlife with the Humane Society of the United States, and the National Wildlife Federation with the National Audubon Society and perhaps increasingly with the World Wildlife Fund.

But the competition between mainstream environmental groups has always been muted and subtle. It does not get in the way of practical coordination on issues, nor cooperation on such matters as employee salary surveys and dealing with regulatory agencies. Where the problem lies is not in this healthy competition, but in the absence of healthy interaction between the more radical groups and the mainstream groups, or even between the pragmatic reformers and the accommodators. Increasingly, the radical groups embody the passion over the issues and articulate the visions of what the future should hold, whereas the mainstream organizations have far more resources and strong management. The dilemma is how to get these two ingredients into a productive relationship. Apart, the radical groups may expend their energy with little tangible results, whereas the mainstream groups may lose their way with no clear vision to pursue.

Challenges Ahead

The challenge of harnessing the resources in the various parts of the environmental movement reminds us of the fact that size and effectiveness are not necessarily correlated. Small organizations can make major impacts, as Stephen Fox pointed out when he analyzed the work of the Emergency Conservation Committee in the 1930s (Fox, 1981). The growth of the major mainstream organizations does not necessarily mean that they are providing efficient ways of improving the environment. It is even arguable that accumulation of organizational resources is counterproductive, in that attention gets focused primarily on that end and on the difficulties of managing complex organizations. Too little time and energy may remain for conducting the work justifying the existence of these organizations—environmental work. Although this result is not inevitable, large management structures make it difficult for fresh ideas to be approved and customary strategies to be reassessed. Too many people have vested interest in the accepted ways, and managers have too little time to think about the need for anything other than management.

Nevertheless, there is a need for the mainstream component of the movement to rethink its assumptions about getting results through government. Its classic experience with getting the government to set aside public land reserves out of the public domain may have misled it. When it got laws passed to do that, they were almost self-executing. Timber is almost never put up for sale within a wilderness area or a national park once it is established, and a lot of oversight is not needed. But these simple experiences with natural resource protection look more and more like the exceptions. Getting regulatory programs for environmental protection implemented is a different matter entirely. They need endless follow-through and can go wrong in a thousand places. The relevant bureaucracies have minds of their own and very little loyalty to the ideas of those who lobbied the programs through. Although the bureaucracies are somewhat responsive to Presidential direction, they are not very responsive to outside lobbying and are subject to no self-correcting process if they fail to be productive (the market does not put them out of business when they are unproductive). Moreover, environmentalists do not seem to be having much success in getting Presidents into power who share their view of the world, and cannot look to the Presidency to rescue them from unresponsive bureaucracies. The pragmatic reformers thus face a crisis in their faith in governmental action.

At the same time, the grassroots are being radicalized by their experiences of receiving no satisfaction from those in power. The government is increasingly perceived as the enemy. They believe it ignores the laws, suppresses evidence, and tells lies. And it certainly does not protect their interests. Why, they may ask, are we advocating giving it more power so that it can keep anything useful from happening? Not only are the local toxics activists feeling this way but so also increasingly are those dealing with public lands issues and wildlife. Local activists across the board feel estranged from the agencies they deal with, whether they be county government or federal agencies. Agencies such as the Forest Service have admitted in their own internal planning reviews that they are rarely pleasing any constituency anymore.

The mainstream movement must face this lack of faith in government head-on. It should ponder the significance of the Alar case in 1989, when NRDC went on national television to denounce use of this chemical on apples. The question about the safety of Alar had been debated for 15 years within EPA, which could not make up its mind about whether to force its withdrawal. After the TV exposé, no customers wanted to buy Alar-treated apples anymore. No supermarkets wanted to sell them. The apple growers pledged to stop using Alar, and the manufacturer of it announced that production was ceasing because the market had collapsed. Even though friends of Alar in industry wanted to continue arguing the case for its safety, the issue had been decided in the marketplace. It did not matter whether EPA banned it; society had decided through other means that its use was unacceptable.

With overwhelming levels of support in public opinion, this kind of action can be repeated. Environmentalists can bypass government in the regulatory field wherever a strong consumer handle exists. Cases can be taken to the public, which can be urged not to buy products produced in damaging ways or containing unsafe ingredients. The most egregious polluters can be the subject of major campaigns to punish them in the marketplace. Boycotts and shareholder actions can become the new stock-in-trade of environmentalists. One should note that those joining in Earth Day events were asked to take the "Green Pledge" to use these tools (Earthworks Group, 1989; MacEachern, 1990; Rifkin, 1990).

The key to success here is information—information on what is in products, what happens in the production processes, and what happens to the product as waste. Fortunately, with the advent of Title III in the new Superfund law, information is becoming available that discloses the names of the companies discharging the greatest amounts of toxics into our air and environment. This information needs to be combined with information on the frequency of violations of pollution and OSHA laws to provide good measures of corporations' environmental records. State laws such as Proposition 65 in California are providing consumers with information about toxics in products. Better labeling laws are needed, too, to arm people with information that will enable them to make their own informed consumer choices.

Tactics along these lines borrow something from the camps at either end of the environmental spectrum.[3] They acknowledge some of the validity of the critiques from the right with respect to the cumbersomeness of government and the need to use market mechanisms. They also acknowledge the frustration of the radicals with respect to government unresponsiveness and the consequent need for direct action. But the mainstream organizations have something to contribute too. Unlike those on the right, they would encourage the use of market mechanisms directly rather than indirectly (as through regulatory mechanisms that mimic markets). Unlike the radicals, they could organize action in this area so that major resources are brought to bear in pursuit of a well-orchestrated national strategy. This could be a serious, long-term effort with staying power and a systematic focus on consumer education. It would not be an impulsive hit-and-run action.

Direct action in the marketplace would be intelligible to constituencies. It does not require learning the jargon of regulators and scientific experts. Specific targets would be pinpointed, with clear sets of objectives. The reasons why the objectives are important would be set forth clearly. This tactic would mobilize latent and younger constituencies alienated from the political process and would cut across all ages and classes.

Direct action in the marketplace and at stockholders' meetings might also help to unify the movement. It would be a strategy that draws on the insights and contributions of all camps and, most importantly, would offer the greatest promise of making major gains in the real world. Quite unexpectedly, it might also tend to relieve some of the pressure on overstressed regulatory agencies such as EPA and FDA. It would not work as well for land management agencies, such as the Forest Service and the Bureau of Land Management, but these agencies are now being forced to find a new modus vivendi with enraged constituencies in the states where they operate. The Forest Service admits that it needs to start pleasing somebody, someplace, some of the time; it cannot remain at war with everyone.

The environmental movement has not lost its commitment or determination, but it is at a crossroads in terms of knowing how to produce results. The breakup into competing camps reflects this uncertainty. Those at the extremes may care less about getting results than about feeling that they are right.[4] However, the pragmatists in between must care about results, or they have forfeited their reason for existing. They now must bear the burden of rethinking their fundamental goals, strategies, and tactics.

Fortunately, there seems to be an answer. Whether they will see it or find another will tell us much about the continuing vitality of the movement.

Notes

1. Many groups participated in running the campaign in the late 1970s for the Alaska Public Interest Lands Act; it was truly a major campaign and was reminiscent of the original Wilderness Act campaign in size and complexity. No campaign in the 1980s came close to it, although the Sierra Club and the Wilderness Society put major resources into the anti-Watt campaign in the early 1980s, and the Sierra Club put significant resources into the battle for the Clean Air Act and the Superfund reauthorization as well as into the defense of the Arctic National Wildlife Refuge.

2. For instance, there were splits with regard to strategies between the National Resources Defense Council and the National Audubon Society over the 1986 Clean Water Act reauthorization.

3. They combine what some have called power and participative strategies and can go beyond boycotts to involve picketing too.

4. And, of course, they remain far apart in goals.

References

Adams, J. H., Dunlap, L. C., Hair, J. D., Krupp, F. D., Lorenz, J., McCloskey, J. M., Peterson, R. W., Pritchard, P. C., Turnage, W. C., and Wendelowski, K. 1985. *An environmental agenda for the future.* Washington, DC: Island Press.

Bookchin, M. 1990. *Remaking society: Pathways to a green future.* Boston: South End Press.

Borrelli, P. ed. 1988. *Crossroads: Environmental priorities for the future.* Washington, DC: Island Press.

Dunlap, R. E., 1987. Public opinion and the environment in the Reagan era. *Environment,* 29:7–11, 32–37.

——— . 1989. Public opinion and environmental policy. In *Environmental politics and policy,* ed. J. P. Lester, pp. 87–134. Durham: Duke University Press.

Earthworks Group. 1989. *50 simple things you can do to save the Earth.* Berkeley, CA: Earthworks Press.

Fox, S. 1981. *John Muir and his legacy: The American conservation movement.* Boston: Little, Brown.

Freudenberg, N. 1984. *Not in our backyards!* New York: Monthly Review Press.

Friends of the Earth. 1982. *Indictment: The case against the Reagan environmental record.* Washington, DC: Author.

Hays, S. 1959. *Conservation and the gospel of efficiency.* Cambridge, MA: Harvard University Press.

Humphrey, C., and Buttel, F. 1982. *Environment, energy, and society.* Belmont, CA: Wadsworth Publishing Company.

MacEachern, D. 1990. *Save our planet: 750 everyday ways you can help clean up our Earth.* New York: Dell Publishing.

Maize, K. ed. 1988. *Blueprint for the environment.* Washington, DC: Blueprint for the Environment.

Morrison, D., Hornbach, K. E., and Warner, W. K. 1972. The environmental movement: Some preliminary observations and predictions. In *Social behavior, natural resources, and the environment,* ed. W. R. Burch, Jr., N. H. Cheek, Jr., and L. Taylor, pp. 259–279. New York: Harper & Row.

Rifkin, J. ed. 1990. *The green lifestyle handbook: 1001 ways you can heal the Earth.* New York: Henry Holt.

Trends in Public Opinion Toward Environmental Issues: 1965–1990

RILEY E. DUNLAP

Departments of Sociology and Rural Sociology
Washington State University
Pullman, WA 99164-4006
USA

Abstract *A comprehensive review of available longitudinal data on public opinion toward environmental issues since 1965 suggests the following conclusions regarding trends in public concern over environmental quality: (a) Environmental concern developed dramatically in the late 1960s and reached a peak with the first Earth Day in 1970; (b) such concern declined considerably in the early 1970s and then more gradually over the rest of the decade, but remained substantial; (c) the 1980s saw a significant and steady increase in both public awareness of the seriousness of environmental problems and in support for environmental protection, with the result that by the twentieth anniversary of Earth Day in 1990, public concern for environmental quality reached unprecedented levels. This supportive public opinion provides a valuable resource for the environmental movement, and the future of the movement will depend heavily on the degree to which environmentalists can effectively mobilize this support.*

Keywords Earth Day, environmental attitudes, environmental legislation, environmental movement, environmental opinions, environmental perceptions, environmentalism, government actions, public opinion.

About a quarter century ago, environmental quality began to emerge as a major social problem in our society. Issues such as wilderness protection and air pollution had previously received the attention of relatively small numbers of conservationists and public health officials, but in the mid 1960s a wide range of threats to environmental quality began to attract the attention of the media, policymakers, and the public. Due in large part to the efforts of a growing number of environmental activists, by 1970 the environment had become a major national concern, as reflected by the huge scale of Earth Day (22 April 1970) celebrations across the nation (see, e.g., Mitchell and Davies, 1978). Over the past two decades, environmentalism has evolved into a major sociopolitical force in our society, and one of the key reasons for this has been the widespread support that the environmental movement has received from the general public.

Social movements are spearheaded by activists and organizations, but their success or failure is often heavily influenced by the degree of support they receive from the broader public. Although contributions of money or time and political support (voting, petition signing, letter writing, etc.) are obviously crucial, the mere expression of sup-

Thanks are due to Angela Mertig and Rik Scarce for helpful comments on earlier drafts.

port by the public in a scientific survey or an informal poll (as are often conducted by local newspapers and politicians) can also be a vital resource for a social movement. As Mitchell noted, "Public support of environmental groups provides them with a key lobbying resource because it lends credibility to the claim that they represent the 'public interest' " (1984, p. 52). Supportive public opinion thus not only lends legitimacy to a social movement but it provides a valuable resource in lobbying for new legislation or pressing for the effective implementation of existing legislation (Rosenbaum, 1991; Sabatier and Mazmanian, 1980). Some analysts even conceptualize the "sympathetic public" as an outer layer of the environmental movement from which the core activists and organizations frequently can recruit members and solicit contributions and other resources (e.g., Morrison, 1986).

For such reasons the status of public opinion on environmental issues has received a great deal of attention over the past two decades. The degree to which the public supports efforts to protect environmental quality, and whether such support has increased or decreased, has been the subject of considerable debate over the years (see, e.g., Ladd, 1982). My goal here is to present a comprehensive overview of the evidence available on *trends* in public concern for environmental quality, beginning with the emergence of such concern in the mid 1960s and continuing through its evolution up to the twentieth anniversary of Earth Day in 1990.

The task is difficult because there are no data sets that have continuously monitored public opinion on environmental issues over this entire time period. However, by piecing together several sets of relevant longitudinal data—covering the late 1960s to 1970, the early 1970s, the mid to late 1970s, and the 1980s—I hope to provide an accurate portrayal of the broad contours of trends in public concern for environmental quality over the past quarter century. However, first I want to discuss a model of the evolution of public opinion toward social problems in general, and environmental problems in particular, in an effort to establish what was expected to happen to public support for environmental protection over this period.

The Natural Decline Model

As noted in the introduction to this volume, social scientists have discerned a pattern in which social problems are regularly discovered or "created" by a group of activists who are successful in getting the larger society to accept their definition of various conditions as problematic and in need of amelioration. Such efforts are frequently transitory and seldom fully successful, however, and generally experience what can be termed a *natural decline*. One major reason seems to be the decline in public interest in and attention to the problem (Dunlap, 1989). Whether from basic boredom with the issue, from the fact that the media turn attention to newer issues, or from the sense that government is taking care of the problem and there is no longer any need to worry about it, the public is seen as inevitably losing interest in most social problems (e.g., Mauss, 1975; Sabatier and Mazmanian, 1980).

Given the thrust of social science thinking about the evolution of social problems and public concern with such problems, it is not surprising that a compatible model developed specifically for environmental problems by Anthony Downs—the *issue-attention cycle*—has been quite influential in analyses of public opinion on environmental issues. Writing a couple of years after the first Earth Day, Downs (1972) suggested that environmental problems would probably meet the fate experienced by most social problems: have a brief moment in the sun and then fade from public attention as newer problems take center stage on the national agenda.

Specifically, Downs (1972) suggested that, like most social problems, environmental problems would proceed through a five-stage cycle:

(1) the *pre-problem* stage in which the undesirable social conditions exist and may have aroused the interest of experts or interest groups but have not yet attracted much attention from the public;
(2) the *alarmed discovery and euphoric enthusiasm* stage in which one or more dramatic events or crises bring the problem to the public's attention and create enthusiastic support for solving it;
(3) a *realization of the cost of significant progress* stage in which public enthusiasm is dampened;
(4) a gradual *decline in the intense public interest* due to recognition of the costs of a solution, boredom with the issue, and decline in media attention to the problem; and finally
(5) the *post-problem* stage in which the issue is replaced at the center of public concern by new problems and moves into "a twilight realm of lesser attention or spasmodic recurrences of interest" (Downs, 1972, p. 40), typically with little if any improvement in the original problematic conditions.[1]

Downs went on to suggest that in 1972 environmental quality was already about halfway through the issue-attention cycle.

Downs's model, which is very much in line with social science models of the evolution of social problems, social movements, and public policy implementation, will be used as a benchmark in reviewing trends in public opinion toward environmental issues. Using it will enable me to determine the degree to which the environmental movement has managed to succeed in maintaining a high level of public attention to environmental problems and concern for environmental protection, thereby avoiding the natural decline in public support that is the fate of the typical, short-lived social-problem movement.

Trends in Public Concern for Environmental Quality

The Early 1960s to 1970: Emergence of Widespread Public Concern

The fact that environmental issues were virtually ignored by public opinion pollsters in the early 1960s indicates the low level of societal attention to such issues at that time and suggests that environmental quality was still in Downs's (1972) pre-problem stage (of interest mainly to conservationists and air and water pollution officials). The situation rapidly changed in the latter half of the decade, however, because of several interrelated developments. First, conservation organizations such as the Sierra Club became more visible, both because they were appealing for widespread support in battles such as the fight to save the Grand Canyon and because they were broadening their focus beyond wilderness and scenic preservation to a wider range of environmental issues (Hays, 1987). Their activities coincided with a rapid increase in outdoor recreation, another factor contributing to the growth in membership of conservation organizations (McEvoy, 1972). At the same time, political leaders such as President Johnson and Senators Muskie and Jackson were pushing important environmental legislation through Congress, ranging from measures to improve air and water quality to protecting endangered species and culminating in the landmark National Environmental Policy Act (NEPA) of late 1969 (Hays, 1987). These trends and activities no doubt helped sensitize the mass media to environmental issues, and by the late 1960s environmental problems were receiving tremendous exposure in the media (Schoenfeld, Meier, and Griffin, 1979). Finally, there was the enormous mobilization of citizen participation, facilitated by the widespread social activism of the 1960s, in celebration of the first Earth Day in the spring of 1970.

The effect of all of this on public opinion toward environmental issues is illustrated by several sets of trend data summarized in Table 1. The first set of data, from Gallup polls conducted in 1965 and 1970, show that the percentage of the public selecting "reducing pollution of air and water" as a national problem that should receive the attention of government more than tripled (from 17% to 53%) during those 5 years. Almost as impressive an increase is documented in a series of Opinion Research Corporation surveys covering the same period, as the percentages viewing air and water pollution as "very or somewhat serious" in their vicinity more than doubled, from 28% to 69% for air pollution and from 35% to 74% for water pollution. A bit less impressive was the increase in the percentage responding that there was "a lot" or "some" air pollution in their area, from 56% to 70%, in 1967 and 1970 Harris surveys, respectively.[2] The Harris surveys also asked respondents if they would be willing to pay $15 per year more in taxes for an air pollution control program, and over the 3 years those responding "willing" rose from 44% to 54%, whereas those saying "unwilling" dropped from 46% to 34%. The last set of Harris data covers only seven months, from August 1969 to March 1970, but reflects the impact of mobilization for Earth Day on public opinion: The percentage selecting "pollution control" as one of the three or four government programs (from a list of ten) that they would "least like to see cut" increased by nearly half, from 38% to 55%.

These trend data, especially those covering 1965 to 1970, indicate how dramatically public concern with environmental quality emerged during the last half of the 1960s. Especially notable is the fact that in the Gallup surveys the 17% selecting pollution reduction ranked it ninth overall among the ten problems in 1965, whereas the 53% selecting it in 1970 placed it second only to crime reduction. Such results led Erskine to conclude, "A miracle of public opinion has been the unprecedented speed and urgency with which ecological issues have burst into American consciousness. Alarm about the environment sprang from nowhere to major proportions in a few short years" (1972, p. 120). The data also suggest that by 1970 environmental quality had definitely moved from the pre-problem stage to the alarmed discovery stage in Downs's (1972) issue-attention cycle, the point at which the public clearly acknowledges the seriousness of a problem and enthusiastically supports efforts to solve it.

In the terminology of public opinion analysts (e.g., Pierce, Beatty, and Hagner, 1982), environmental protection had become a consensual issue by 1970, as surveys found a majority of the public expressing pro-environment opinions and typically only a small minority expressing anti-environment opinions. The high and consensual level of concern for environmental quality at the beginning of what would become known as the "environmental decade" was also reflected by cross-sectional data collected in 1969 to 1970 (Erskine, 1972). Yet, was environmental quality really that salient as an issue, one that was truly on the public's mind, even during this period of strong support for environmental protection? This is difficult to judge, but many public opinion analysts have argued that volunteered responses to most important problem (MIP) questions—that is,

Table 1

Trends in Public Concern for Environmental Quality, Mid 1960s to 1970

National Survey	Question	Percentage Response by Year (19—)					
		65	66	67	68	69	70
(1) Gallup	''Reducing pollution of air and water'' selected as one of three national problems that should receive attention of government	17	—	—	—	—	53
(2) Opinion Research Corporation	Air/water pollution viewed as ''very or somewhat serious'' in the area:						
	(a) air pollution	28	48	53	55	—	69
	(b) water pollution	35	49	52	58	—	74
(3) Louis Harris	''A lot'' or ''some'' air pollution thought to exist in the area			56	—	—	70
(4) Louis Harris	Willing to pay $15 a year more in taxes to finance air pollution control program			44	—	—	54
(5) Louis Harris	''Pollution control'' selected as government spending area ''least like to see cut''					38	55

Note. For each survey, see the following reference for question wording and complete results: (1) Mitchell (1980, p. 404); (2) Erskine (1972, p. 121); (3) Erskine (1972, p. 123); (4) Erskine (1972, p. 132); (5) Erskine (1972, p. 129).

open-ended questions asking respondents what they see as the country's most important problem or problems—are the best way to measure the salience of an issue (e.g., Peters and Hogwood, 1985). MIP questions are stringent measures of salience, however, because responses are traditionally dominated by economic issues and foreign affairs/ national security concerns (Smith, 1985).[3]

Unfortunately, MIP studies that report results for environmental problems are rare, and because the only two MIP trend studies that began in the 1960s extended past 1970, I have summarized them in Table 2 along with other longitudinal studies beginning in 1970. The sole trend study of the salience of environmental quality at the national level beginning in the 1960s was Hornback's (1974) analysis of MIP data collected in the Michigan National Election Surveys (NES), in which respondents were encouraged (through probes) to mention up to three problems facing the country. Hornback found that only 2% of the public volunteered any type of environmental problem in 1968, a surprisingly low figure in view of the data reported in Table 1 for other indicators of concern about environmental quality. In 1970 the figure was 17%, representing a dramatic increase but still only a small proportion of the public. The other MIP study from the 1960s was limited to the state of Wisconsin, in which Buttel (1975a, 1975b) found that the percentage of residents volunteering environmental problems as one of the two most important problems facing the state rose substantially, from 17% to 40% between 1968 and 1970. Although the degree of increase in Wisconsin was similar to that for the nation, the absolute levels of salience in Wisconsin were much higher in both years. Whether this reflects a higher than average level of environmental concern in Wisconsin or differences in study methodologies (or both) is impossible to determine.

Matters are not helped when the 1970 data from these two studies are compared with two other sets of MIP data beginning in 1970. First, Harris began reporting the results of an MIP question in late 1970. As shown in Table 2, fully 41% of a national sample volunteered some type of environmental problem as one of the "two or three biggest problems" facing them in 1970—a figure comparable with the Wisconsin results. In contrast, three 1970 Gallup surveys included a question asking about the single "most important problem" facing the country, and the percentages mentioning environmental problems ranged from only 2% to 10%. The Gallup results, taking into account that they asked for only one MIP, thus seem more in line with the results from the Michigan NES reported by Hornback.

The discrepant results obtained with MIP questions, especially those between the Michigan NES and Harris surveys in 1970, are difficult to reconcile. They do not seem attributable, for example, to differences in question wording. Despite the inconsistency in results, however, two conclusions can be drawn from the available MIP data. First, the salience of environmental problems increased substantially from 1968 to 1970. Second, even in 1970 only a minority (albeit a large one in some surveys) volunteered environmental problems when asked what they saw as the most important problems facing the country. The latter finding contrasts sharply with the results reported in Table 1 for other indicators of public concern for environmental quality and suggests that even around the time that our nation was celebrating the first Earth Day the salience of environmental problems did not match that of the traditionally dominant worries about war and the economy (Erskine, 1972; Hornback, 1974; Smith, 1985).

Overall, therefore, the available data indicate that public concern for environmental quality escalated rapidly in the 1960s and that by 1970 majorities of the public were expressing pro-environment opinions ranging from acknowledging the seriousness of

Table 2

Longitudinal Studies of Public Concern for Environmental Quality, Late 1960s to Mid 1970s

Study	Question	Percentage Response by Year (19—)								
		68	69	70	71	72	73	74	75	76
National Survey										
(1) Michigan National Election Survey	Pollution, ecology, etc., volunteered as one of the country's "most important problems"	2	—	17	—	10				
(2) Louis Harris	Pollution, ecology, etc., volunteered as one of "the two or three biggest problems facing people like yourself"			41	—	13	11	9	6	
State Trend/Panel Studies										
(3) Wisconsin	Environmental problems volunteered as one of two most important facing the state	17	—	40	—	15	—	10		
(4) Washington (panel)	Favor government spending "more money" on:									
	(a) pollution control			70	—	—	—	32		
	(b) protection of natural resources			52	—	—	—	37		
(5) Washington (trend)	"Reducing air and water pollution" selected one of two or three most serious problems in:									
	(a) state			44	—	—	—	—	—	18
	(b) respondents' community			23	—	—	—	—	—	15

Note. For each survey, see the following reference for question wording, complete results, and response coding: (1) Hornback (1974, pp. 87, 233–234); (2) Mitchell and Davies (1978, Figure 2); (3) Buttel (1975a, pp. 83–85; 1975b, p. 58); (4) Dunlap and Dillman (1976, p. 383–384); (5) Dunlap and Van Liere (1977, p. 110).

pollution to supporting governmental efforts to protect and improve the environment. However, despite the relatively strong consensus in support of environmental protection, the state of the environment was viewed by only a minority of the public as one of the nation's most important problems. From a social movements perspective, it appears that a majority of the public had accepted environmentalists' definition of environmental quality as problematic and had become sympathetic to the goal of protecting the environment, but only a minority saw the environment as one of the nation's most important problems.

The Early 1970s

As the high level of environmental activism stimulated by Earth Day inevitably began to decline, several commentators suggested that public support for environmental protection would likewise decline. The most influential was Downs (1972), who (as noted earlier) suggested that by 1972 environmental problems had already passed from the stage of alarmed discovery and enthusiastic support into one of somber realization of the costs of environmental protection and improvement. The situation certainly seemed conducive for environmental quality to pass through the issue-attention cycle. At the beginning of the decade the government (both federal and state) passed a great deal of environmental legislation, set up highly visible environmental agencies (most notably the Environmental Protection Agency), and spent a good deal of money on behalf of environmental improvement and protection (Lester, 1989), thus giving the impression that government was taking care of environmental problems. In addition, media attention to environmental problems began to decline after 1970, and such problems were eclipsed when the energy crisis of 1973–1974 took over center stage on the public agenda (Schoenfeld et al., 1979). The setting seemed ripe for environmental problems to pass through the final stages of Downs's cycle and experience the natural decline posited by social scientists.

What, in fact, happened to public concern for environmental quality in the early 1970s? Sadly, data needed for providing a reasonably definitive answer to this question are not available. Not only did pollsters often stop asking the environmental questions they had used in the 1960s, but, surprisingly, they failed to start asking new questions in 1970 to provide a baseline for monitoring changes in environmental concern. The situation led Erskine (1972) to express consternation over public opinion pollsters' failure to collect good trend data on environmental issues, and forces one to rely on the very limited body of data shown in Table 2.

I have already referred to three of the data sets reported in Table 2, both of the national studies and the Wisconsin study. All three used MIP questions and all show a similar pattern: The salience of environmental problems declined substantially by 1972 from its peak in 1970, and even further by 1974–1975. The patterns in the Harris and Wisconsin data are especially similar, and indicate that the proportion of the public volunteering environmental problems as among the nation's or state's most serious problems declined from a large minority in 1970 to a small minority (10% or less) by mid-decade. A similar pattern was found in Gallup surveys using an MIP question asking for the *single* most important problem. As noted earlier, the percentage volunteering environmental problems reached a peak of 10% in 1970, and then fluctuated between 7% and 2% during 1971 and 1972 (Hornback, 1974). Unfortunately, in 1973 Gallup began asking respondents to name the *two* MIPs, making comparisons with prior years impossible. The environment continued to show up low on the Gallup MIP lists through 1973 but dropped off in 1974, when it was replaced by energy (Smith, 1985).[4]

Earlier I noted that public opinion analysts tend to regard responses to MIP questions as good indicators of the salience of an issue to the public. But it has been argued that responses to such questions are especially susceptible to media attention to particular problems (Funkhouser, 1973) and that—more broadly—salience as measured by MIP responses "is transitory for all but the most momentous issues such as war or depression" (Mitchell, 1984, p. 55). Because media attention to environmental problems declined considerably in the early 1970s (Schoenfeld et al., 1979), perhaps these trends reflect little more than the public's susceptibility to the agenda-setting function of the mass media. The results also raise questions about the validity of MIP questions for measuring issue salience.[5] Some analysts of environmental concern have, in fact, argued that the obvious decline in the salience of environmental problems in the early 1970s was not matched by a decline in public commitment to environmental protection (Mitchell, 1984).

Remarkably, the only two sets of data available for testing this possibility are limited to the state of Washington. Fortunately, although the two studies used very different indicators of concern for environmental quality, they nonetheless produced similar results. In both cases a fairly sharp decline in environmental commitment was found by mid-decade, although not as great as the decline shown by the MIP indicators. The 1970–1974 panel study compared the priorities for spending government funds for 1600+ Washington residents over the 4 years and found significant declines in the percentage wanting more tax money spent on environmental protection. Those wanting more spending on pollution control declined from 70% to 32%, whereas those wanting more spent on protection of forests and other natural areas for public enjoyment declined from 52% to 37%. The 1970–1976 trend study compared different samples of 800+ residents in terms of their selection of reducing air and water pollution as one of the two or three most serious problems facing the state and their communities (out of a list of eleven potential problems). At the state level there was a large decline, from 44 to 18 in the percentage selecting "reducing pollution" over the 6 years. The decline at the community level was smaller, from 23% to 15%, due in part to the fact that respondents were much less likely to see pollution as a community problem to begin with (Dunlap and Van Liere, 1977).

In short, the two Washington State studies found that public concern for environmental quality, measured both by spending priorities and by perceived seriousness of environmental problems, declined substantially between 1970 and mid-decade. Especially noteworthy is that in the 1970–1974 panel study, not only did the percentage wanting more spending on pollution control decline, but the percentage wanting less spending in this area increased as well (from 5% to 21%). Although great caution is called for in generalizing from a single state, the Washington results reveal a deterioration of the strong consensus on behalf of environmental protection that emerged with the first Earth Day in 1970.

To summarize, although the available evidence on trends in public concern for environmental quality in the early 1970s is sparse, it consistently indicates a significant decline in environmental awareness and concern among the public in the early 1970s. This seems to support Downs's (1972) contention that by 1972 environmental quality was about halfway through the issue-attention cycle, and his prediction that it would shortly move into the fourth stage (the decline of intense public interest) consequently seemed very plausible by mid-decade (see, e.g., Dunlap and Dillman, 1976).

Table 3
Trends in Public Concern for Environmental Quality, Early 1970s to 1980

National Survey	Question	Percentage Response by Year (19—)							
		73	74	75	76	77	78	79	80
(1) Roper[a]	More on the side of:								
	(a) protecting the environment	37	39	39	44	35	—	38	36
	(b) having adequate energy	37	41	40	33	43	—	43	45
(2) Roper[b]	Environmental protection laws and regulations have gone:								
	(a) not far enough	34	25	31	32	27	—	29	33
	(b) too far	13	17	20	15	20	—	24	25
(3) NORC[c]	U.S. spending on improving and protecting the environment:								
	(a) too little	61	59	53	55	48	52	—	48
	(b) too much	7	8	10	9	11	10	—	15
(4) Cambridge[d]	Sacrifice environmental quality or sacrifice economic growth:								
	(a) sacrifice economic growth				38	39	37	37	—
	(b) sacrifice environmental quality				21	26	23	32	—
(5) Roper[e]	Will be a "serious problem" 25 to 50 years from now:								
	(a) severe air pollution		68						68
	(b) severe water pollution		69						69
	(c) shortage of water supplies		53						57

Note. NORC = National Opinion Research Center.

[a]Full question: ‘‘There is continuing talk about an energy crisis and the idea that there won’t be enough electricity and other forms of energy to meet consumer demand in the coming years. Some people say that the progress of this nation depends on an adequate supply of energy and we have to have it even though it means taking some risks with the environment. Others say the important thing is the environment, and that it is better to risk not having enough energy than to risk spoiling our environment. Are you more on the side of adequate energy or more on the side of protecting the environment?’’ Volunteered responses of ‘‘neither,’’ ‘‘no conflict,’’ or ‘‘don’t know’’ are not shown. Results are reported in Gillroy and Shapiro (1986, p. 275) and Dunlap (1987, p. 8).

[b]Full question: ‘‘There are also different opinions about how far we’ve gone with environmental protection laws and regulations. At the present time, do you think environmental protection laws and regulations have gone too far, or not far enough, or have struck about the right balance?’’ Percentages responding ‘‘struck about right balance’’ or volunteering ‘‘don’t know’’ are not shown. Results are reported in Gillroy and Shapiro (1986, p. 273) and Dunlap (1987, p. 9).

[c]Full question: ‘‘We are faced with many problems in this country, none of which can be solved easily or inexpensively. I’m going to name some of these problems, and for each one I’d like you to tell me whether you think we’re spending too much money on it, too little money, or about the right amount. First . . . Are we spending too much, too little, or about the right amount on . . . Improving and protecting the environment?’’ Percentages responding ‘‘about right’’ or volunteering ‘‘don’t know’’ are not shown. Results are reported in National Opinion Research Center (1989, pp. 104, 108).

[d]Full question: ‘‘Which of these two statements is closer to your opinion: We must be prepared to sacrifice environmental quality for economic growth. We must sacrifice economic growth in order to preserve and protect the environment.’’ Percentages volunteering ‘‘don’t know’’ are not shown. Results through 1986 are reported in Cambridge Reports, Inc. (1986, p. 9) and Dunlap (1987, p. 11).

[e]Full question: ‘‘Here is a list of some different kinds of problems people might be facing 25 to 50 years from now. Would you please go down that list and for each one tell me whether you think it will be a serious problem your children or grandchildren will be facing 25 to 50 years from now?’’ Results are reported in Roper’s *The Public Pulse,* (New York, NY), ‘‘Research Supplement,’’ June 1989, p. 1.

The Mid to Late 1970s

Although a few years late, by 1973 three items measuring public concern for environmental quality began to be used on a regular basis in national surveys, and more were added later in the decade. In reviewing the results obtained with these items (shown in Table 3), it must be kept in mind that by the time they were first used, public concern for the environment had already declined significantly from its peak in 1970.

In late 1973 (reflecting the emergence of the energy crisis), Roper began using a trade-off question in which respondents were asked if they were more on the side of producing adequate energy or more on the side of protecting the environment. For the sake of brevity I have reported only the percentages for these two positions, deleting the sizable proportion of respondents who either volunteered "neither" or "no conflict" or indicated "don't know." In 1973 the two positions received equal levels of support (37% each), but after that the percentage siding with adequate energy began to exceed that siding with environmental protection (with the exception of 1976). By 1980 the gap had reached a modest 9% (45% vs. 36%), indicating that worries about energy supplies had clearly exceeded concern about environmental protection—but had not caused an erosion of support for environmental protection as was widely expected, for the latter held virtually constant from 1973 to 1980 (from 37% to 36%).

The second Roper question, asking respondents whether they think environmental protection laws and regulations have gone too far, or not far enough, or have struck about the right balance, shows a more substantial decline in public support for environmental protection. Table 3 shows the percentages indicating "not far enough" or "too far" (the percentages indicating "struck about the right balance" or "don't know" were deleted for brevity). In 1973 the percentage indicating that environmental protection efforts had not gone far enough sharply exceeded that indicating that such efforts had gone too far (34% vs. 13%). But the gap quickly began to close (with the exception of 1976) and reached a low point of only 5% in 1979 (29% vs. 24%).

The next item is from a question used by the National Opinion Research Center (NORC) in which respondents are given a long list of problems facing our nation and asked if they think we're spending "too much money on it," "too little money," or "about the right amount" for each one. Surprisingly, as late as 1973 the percentage responding that "too little" was being spent on improving and protecting the environment overwhelmed the percentage responding that "too much" was being spent in this area: 61% versus 7% (the percentages indicating "about right" and "don't know" are not shown). In subsequent years there was a modest but fairly consistent decline in the percentage responding "too little," and a small but relatively consistent increase in the percentage indicating "too much." The result is that the initial 54 percentage points difference between these two positions in 1973 declined to 33 percentage points by the end of the decade (National Opinion Research Center, 1989).

The fourth item, which Cambridge did not use until 1976, poses a broad trade-off between economic growth and environmental quality. It asks respondents whether we must sacrifice economic growth in order to preserve and protect the environment or sacrifice environmental quality for economic growth. In 1976 the public was almost twice as likely to prefer sacrificing economic growth for environmental quality as vice versa (38% vs 21%). Although the percentage choosing to sacrifice economic growth held nearly constant over the next 3 years, the percentage indicating a willingness to sacrifice environmental quality steadily grew (with a consequent decline in the large proportion of volunteered "don't knows," which are not shown). The result is that by

1979 the pro-environment position had only a 5 point margin over the pro-growth position (37% vs. 32%).

The final set of data in Table 3 covers only two points in time and is somewhat at odds with the other results in this table, and thus it should be viewed with caution. Nonetheless, the 1974 and 1980 Roper data on perceptions of the degree to which various environmental problems will probably be "serious problems" in the future (25 to 50 years) reveal a remarkable degree of stability, with the percentages viewing air and water pollution as likely to be serious remaining identical (68% and 69%, respectively) and the percentages viewing water shortages as probable increasing slightly (from 53% to 57%) over the 6 years. The fact that majorities of the public saw pollution and water shortages as future problems suggests that although the saliency of environmental problems declined substantially and support for environmental protection declined modestly during the 1970s, the public apparently did not see environmental problems as disappearing.

The data covering the 1970s in Table 3 provide a generally consistent image of trends in environmental concern from the early part of the mid 1970s through the end of the decade: modest but continued decline in public support for environmental protection. These results, coupled with those reviewed in Tables 1 and 2, suggest the following evolution of public concern with environmental quality: Concern grew rapidly in the late 1960s, reached a peak in 1970 after Earth Day, experienced a fairly sharp decline in the early part of the 1970s, and declined gradually throughout the rest of the decade.

The long-term trend I have just sketched is supported by one final piece of longitudinal data (covering too long a time span to be summarized in the prior tables). A 1980 survey for the Council on Environmental Quality (CEQ) repeated the question used by Gallup in 1965 and 1970 (listed in Table 1). Recall that respondents were given a list of ten national problems, including reducing pollution of air and water, and were asked to indicate which three they wanted to see government devote most of its attention to in the next year or two. As noted previously, reducing pollution was selected by only 17% in 1965, ranking it ninth out of the ten problems; in 1970 it was selected by 53%, hiking it to second. A decade later, in the 1980 CEQ survey, reducing pollution was chosen by 24%, ranking it sixth out of the same list of ten problems (Mitchell, 1980).[6]

In view of these data, should we conclude that at the end of the "environmental decade," public concern for environmental quality had declined to the point that environmental issues had moved into the post-problem stage of the issue-attention cycle? This is not an easy question to answer, largely because Downs vaguely defines the final stage as when "an issue that has been replaced at the center of public concern moves into a prolonged limbo—a twilight realm of lesser attention or spasmodic recurrences of interest" (1972, p. 40). It is clear that if one were to judge public concern by salience, as measured by MIP responses, then one should conclude that environmental quality had moved into the post-problem stage—indeed, it apparently did so as early as 1974 when it was supplanted by the energy crisis. However, the fact that the data in Table 3 indicate at least moderate levels of public concern with environmental quality throughout the decade, long after environment had disappeared from the list of the two or three most frequent responses to MIP questions, again suggests that salience may be a poor indicator of public concern for social problems.[7] Although it is the aspect of public opinion that best reflects the issue-attention cycle, salience may well be unduly influenced by mass media coverage, as argued by Funkhouser (1973).

Even if one ignores the salience dimension, it is difficult to determine if environmental concern had declined to the post-problem stage at the end of the decade.[8] First, note that Downs refers to this stage as a "realm of lesser attention"; second, he later adds that "problems that have gone through the cycle almost always receive a higher average level of attention, public effort, and general concern than those still in the pre-discovery stage" (1972, p. 41). The data reviewed thus far, indicating that environmental concern was lower at the end of the 1970s than at the beginning of the decade, but still higher than in the mid-1960s, would seem to indicate that environmental quality had at least settled into Downs's fourth stage, a period of less intense public interest, by the late 1970s (see, e.g., Anthony, 1982).

Regardless of what one concludes about the issue-attention cycle, it is apparent from Table 3 and from a large amount of cross-sectional data collected in the late 1970s (see Mitchell, 1980) that although public concern for environmental quality had become less consensual, it had by no means disappeared from the public agenda by the end of the 1970s. Although it is clear that the data can be interpreted differently, depending on whether one emphasizes the decline or the endurance of environmental concern throughout the decade, I would generally agree with Mitchell's assessment of the situation in 1980: "Although the state of the environment is no longer viewed as a crisis issue, strong support for environmental protection continues. . . . [F]ar from being a fad, the enthusiasm for environmental improvement which arose in the early 1970s has become a continuing concern" (1980, p. 423). Indeed, I think that public concern for environmental quality showed impressive staying power in the face of a continuing series of essentially competing concerns: the energy crisis of 1973–1974 and continuing concerns about energy supplies throughout the decade, a worsening economic situation, and a taxpayers' revolt begun by California's Proposition 13 in 1978 (Mitchell, 1984). In addition, the fact that the Carter administration was viewed as strongly committed to environmental protection may have contributed to public apathy about environmental problems.[9]

In sum, the environmental movement clearly seemed to have lost some of its broad-based public support throughout the 1970s. Yet, a full decade after the immensely successful Earth Day, and after a vast amount of government action on behalf of environmental quality, the movement had certainly not seen its concerns and goals fade totally from public attention. Public support may have waned, or experienced a natural decline, but it had definitely not disappeared.

The 1980s and the Rejuvenation of Environmentalism

The public may have understandably assumed that government was taking care of environmental problems during the 1970s after watching so much governmental activity in the area, and this assumption was probably a contributing factor to the slow decline in public concern for environmental quality throughout the decade (Dunlap, 1989). However, the situation changed considerably when Ronald Reagan took office in 1980.

Environmentalists were wary of President Reagan because of his general emphasis on deregulation and his tendency to view environmental regulations in particular as hampering the economy (Holden, 1980). The Reagan administration quickly exceeded environmentalists' worst fears, deviating from a decade of generally bipartisan commitment to federal environmental protection. The Council on Environmental Quality was virtually dismantled, the budget of the Environmental Protection Agency (EPA) was severely cut, and the enforcement of environmental regulations was curtailed by administrative review, budgetary restrictions, and staff change. The last category received the most attention, with Anne Gorsuch at EPA and James Watt in the Department of Interior symbolizing the administration's commitment to changing the thrust of environmental policy (Hays, 1987; Portney, 1984; Vig and Kraft, 1984).

It is understandable that environmentalists were upset by the administration's policies and began to vigorously oppose and criticize them. Perhaps most notable was the issuance in 1982 of a well-publicized report, *Ronald Reagan and the American environment,* prepared by ten of the nation's largest environmental organizations and termed an indictment of the administration's environmental policies (Friends of the Earth, 1982). Opposition to these policies grew in Congress, where efforts were made to restore budget cuts and oversee effective enforcement of regulations. Most significant was the Congressional investigation of the EPA's handling of Superfund, which led to the resignation of Gorsuch and several other EPA administrators. Congressional criticism, along with public pressure, also led to the resignation of James Watt, although an ethnic slur rather than the Department of Interior's policies was the precipitating event (see Rosenbaum, 1991).

In the face of mounting criticism, the administration defended its environmental initiatives in terms of its electoral mandate, arguing that Reagan's landslide victory was evidence of the voters' approval of his efforts to free the economy of the burden of governmental regulations (Mitchell, 1984). The President was a vigorous spokesman for deregulation and made the issue a test of his leadership capabilities. Because political leaders have long been recognized as potent forces in shaping public opinion (Pierce et al., 1982), it might be expected that a popular president would have succeeded in convincing the U.S. public that environmental regulations had gone too far. Was President Reagan able to turn the tide against what his administration often termed *environmental extremism*?

The six sets of trend data on public support for environmental quality reported in Table 4 suggest that Reagan was not at all successful in lowering the public's commitment to environmental protection. Indeed, quite the contrary seems to have occurred. In each case there is a pattern of increasing commitment to environmental protection during the Reagan administration, often followed by further increase during the first 2 years of the Bush administration.

The first four items are repeated from Table 3 because they all had been used during the last half of the 1970s. The first poses the trade-off between environmental protection and adequate energy; there was a 10% increase in those siding with the environment and a comparable decrease in those siding with energy from 1980 to 1982 (when Roper temporarily stopped using it). The 46% favoring environmental protection in 1982 was the highest figure recorded with this item, and the 11% margin it enjoyed over energy adequacy matches the previous high point of 1976. These results are especially impressive because a major theme of the Reagan administration was that environmental regulations had to be relaxed to allow for the increased energy production required for a strong economic recovery (Portney, 1984; Vig and Kraft, 1984). Roper began using the item again in 1989, by which time a majority chose environmental protection over energy adequacy, giving the former more than a two-to-one advantage (52% vs. 24%).

The second item, which asks respondents if they think environmental protection laws have gone too far or not far enough, provides an even more direct indicator of the public's evaluation of the Reagan administration's environmental policy agenda. The results indicate a growing rejection of the administration's position: After having reached a low point in 1979, the margin between those indicating not far enough and those indicating too far increased in Reagan's first year in office and in each subsequent year through 1983. In 1982 the 37% saying not far enough already exceeded the pre-

Table 4

Trends in Public Concern for Environmental Quality, 1980 to 1990

National Survey	Question	Percentage Response by Year (19—)										
		80	81	82	83	84	85	86	87	88	89	90
(1) Roper[a]	More on the side of:											
	(a) protecting the environment	36	40	46	—	—	—	—	—	—	57	52
	(b) having adequate energy	45	39	35	—	—	—	—	—	—	24	24
(2) Roper[b]	Environmental protection laws and regulations have gone:											
	(a) not far enough	33	31	37	48	—	—	—	—	—	55	54
	(b) too far	25	21	16	14	—	—	—	—	—	11	11
(3) NORC[c]	U.S. spending on improving and protecting the environment:											
	(a) too little	48	—	50	54	58	58	58	61	65	70	71
	(b) too much	15	—	12	8	7	8	6	6	5	4	4
(4) Cambridge[d]	Sacrifice environmental quality or sacrifice economic growth:											
	(a) sacrifice economic growth		41	41	42	42	53	58	57	52	52	64
	(b) sacrifice environmental quality		26	31	16	27	23	19	23	19	21	15
(5) NYT/CBS[e]	Environmental improvements must be made regardless of cost:											
	(a) agree		45	52	58	—	—	66	—	65	74	74
	(b) disagree		42	41	34	—	—	27	—	22	18	21
(6) Cambridge[f]	Amount of environmental protection by government:											
	(a) too little			35	44	56	54	59	49	53	58	62
	(b) too much			11	9	8	10	7	12	12	9	16

Note. NORC = National Opinion Research Center, NYT/CBS = *New York Times*/Columbia Broadcasting System.

[a]See footnote a of Table 3 for full question. Results for 1989 and 1990 were provided by the Roper Organization.

[b]See footnote b of Table 3 for full question. Results for 1989 and 1990 were provided by the Roper Organization.

[c]See footnote c of Table 3 for full question.

[d]See footnote d of Table 3 for full question. Results through 1989 are reported in Cambridge Reports, Inc. (1989, p. 14). Results for 1990 were provided by Cambridge Reports/Research International.

[e]Full question: "Do you agree or disagree with the following statement: Protecting the environment is so important that requirements and standards cannot be too high, and continuing environmental improvements must be made regardless of cost." Percentages volunteering "no opinion" are not shown. Results through 1989 are reported in *The Polling Report*, 24 April 1989, p. 3. Results for 1990 were provided by the *New York Times*.

[f]Full question: "In general, do you think there is too much, too little, or about the right amount of government regulation and involvement in the area of environmental protection?" Percentages responding "about the right amount" or volunteering "don't know" are not shown. Results through 1989 are reported in Cambridge Reports, Inc. (1989, p. 16). Results for 1990 were provided by Cambridge Reports/Research International.

vious high of 34% from 1973, and the next year it jumped to 48%. Results from the end of the decade indicate that those wanting stronger rather than weaker environmental regulations outnumbered their counterparts five-to-one (54% vs. 11%)!

The next item, the NORC spending item, is also pertinent for judging the impact of Reagan's environmental policy because budget cuts for environmental protection agencies were a major aspect of that policy (Vig and Kraft, 1984). Although still fairly strong, support for increased spending on environmental protection had reached a low point in 1980, with 33% more people indicating "too little" was being spent on the area than indicating "too much" (48% vs. 15%). The gap between these two positions had begun to increase by 1982, and in 1984 it reached 51%, matching the 1974 level and approximating the previous high of 54% in 1973. The gap remained remarkably stable through 1986 and then increased steadily in each of the next 4 years to reach 67% in 1990—well in excess of the 1973 level.[10]

The fourth item, from Cambridge, forces respondents to choose between economic growth and environmental quality. Recall that in 1979, those preferring that economic growth be sacrificed for environmental quality held only a slim margin over those preferring the opposite (37% vs. 32%). At the outset of the Reagan administration the percentage preferring that economic growth be sacrificed rose to 41%, a new high, whereas the percentage opting to sacrifice environmental quality dropped to 26%, producing a margin that nearly equalled the 1976 level when the item was first used. Despite yearly fluctuations, the margin was the same in 1984 and then climbed to new peaks in 1986 (58% vs. 19%) and in 1990 (64% vs. 15%).

The *New York Times*/CBS News Poll began using the next item in September 1981, by which time the Reagan administration was under attack from environmentalists but before its environmental policies had gained the intense media attention of the EPA and Watt controversies. This item asks respondents to react to the extreme pro-environmental position that environmental improvements should be pursued regardless of cost. The public was almost evenly divided on the issue in 1981 (45% vs. 42%), but preference for the pro-environmental position over the rather mild anti-environmental position grew steadily throughout the Reagan years and then took a very large jump in 1989 to reach 56% (74% vs. 18%) and then held ground in 1990.

The final item in Table 4 was not used by Cambridge until March 1982. Like the second Roper item, it provides a good indicator of public reaction to the Reagan administration's overall environmental policy agenda, asking respondents if they think "there is too much, too little, or about the right amount of government regulation and involvement in the area of environmental protection." From the outset, the public clearly rejected the administration's contention that environmental regulations were excessive (by a margin of 35% to 11%), and this view has become more pronounced over time—peaking in 1986 (59% vs. 7%), dipping a bit the next 2 years, and then coming back strong in 1989 and 1990.

Taken together, the six sets of trend data in Table 4 provide a generally consistent view of recent trends in public support for environmental protection. After having declined moderately in the 1970s, public support for environmental protection began to rise shortly after Reagan took office and has continued to do so. This conclusion is bolstered by results obtained with several other items used at two time points only between 1980 and 1990 and reported elsewhere (Dunlap and Scarce, 1991). The trend data, along with a wide range of recent cross-sectional data, thus suggest that environmental protection has again become a consensual issue commanding support from an overwhelming majority of the public and eliciting opposition from only a very small minority. Perhaps most striking is a recent Harris survey that found an amazing 97% responding "more" when asked, "Do you think this country should be doing more or less than it does now to protect the environment and curb pollution?" (Harris, 1989, p. 3).

In staging this comeback, environmental issues have obviously halted their slide into the last stage of Downs's issue-attention cycle, the post-problem stage, and have reversed the natural decline that was the expected course for social problems. This is best illustrated by the only data I have located on environmental concern spanning the entire 1970–1990 period. In both years Gallup used the following item: "As you may know, it will cost a considerable amount of money to control pollution. Would you be willing to pay the slightly higher prices for your goods and services business would have to charge to control pollution?" The proportion responding "yes" increased from 63% to 79%, whereas that answering "no" declined from 27% to 17%, over the two decades ("don't knows" dropped from 10% to 4%).[11] The results, along with other recent survey data (see Dunlap and Scarce, 1991), suggest that public concern with environmental problems is stronger now than it was in 1970.

What has accounted for this rejuvenation of environmental concern in our society? I have argued elsewhere (Dunlap, 1987, 1989) that much of the increased support for environmental protection in the 1980s probably stemmed from the public's apprehension that, unlike its predecessors, the Reagan administration could not be trusted to protect the environment (a perception that was obviously fueled by environmental organizations and the media, with the unwitting support of Watt, Gorsuch, and others). Indeed, large numbers of people became sufficiently concerned that they joined environmental organizations for the first time, producing sizable membership gains for many of the national organizations in the 1980s (see Mitchell, Mertig, and Dunlap in this issue). This interpretation is strengthened by considerable evidence that the public was aware of the administration's poor environmental record and that, in general, the public believes that the government *should* assume responsibility for environmental protection (Dunlap, 1989).

What is notable about the growth of environmental concern in the 1980s, however, is that this concern did not decline after the Reagan administration's environmental scandals receded and that it has continued to climb since Reagan left office. Although President Bush has by no means proven himself to be a strong environmentalist, his efforts to portray himself as concerned about environmental quality have thus far shielded him from the intense criticism that environmentalists leveled at Reagan. This suggests that the other major factor stimulating rising public concern over environmental quality—increased awareness of the growing seriousness of ecological problems (Dunlap, 1989; Mitchell, 1984, 1990)—has probably become the critical force in driving public opinion.

Public Perceptions of Ecological Problems in the 1980s

A combination of what might be called the institutionalized environmental movement (a collection of actors including government officials in environmental agencies; environmental scientists in government, academic, and other nonprofit research centers; and environmental organizations and activists; see Morrison, 1986), a sympathetic mass media, and ecological realities have combined to generate enormous societal attention to ecological problems in recent years. Scientists and government officials are constantly joining environmentalists in publicizing the latest aspects of ecological degradation, and their efforts have been validated by an endless array of newsworthy events (Bhopal, Chernobyl, frequent chemical spills, hazardous ocean beaches, oil spills, rainforest destruction, filled-up waste sites) that receive tremendous media attention. The success of these efforts in attracting public attention can be seen in newspapers, on TV news programs, and on the covers of our nation's most important news magazines (see, e.g., Mitchell, 1990). During 1988–1989 alone, *Newsweek* carried numerous cover stories on ecological problems, *U.S. News and World Report* had one on "Planet Earth: How it works, how to fix it" (31 Oct. 1988), and *Time* captured the most attention by naming the "Endangered Earth" as "Planet of the Year" in lieu of its famous "Man of the Year" for 1988 (2 January, 1989). Finally, there was the enormous amount of media coverage (radio and television as well as print) of the 22 April 1990 celebration of the twentieth Earth Day (Nixon, 1991).

Increase in Perceived Seriousness of Problems

The impact of all of this on public awareness of what have become "ecological" (rather than "environmental") problems is aptly demonstrated by the range of trend data reported in Table 5. Unlike Table 4, in which the items focused on support for environmental protection, the items in Table 5 focus on perceptions of the seriousness of environmental problems. The results from these seven sets of trend data indicate a substantial rise in the public's perception of environmental problems as serious issues during the 1980s.

The first set of Roper data continues the results reported for 1974–1980 in Table 3 for three of the problems and shows sizable increases in the percentages of the public viewing air and water pollution as likely to be "serious problems" 25 to 50 years in the future (14% and 13%, respectively, from 1980 to 1988). There were also substantial increases of 9% for both water shortages and overpopulation (included by Roper since 1980). Most striking is the increase of 28% for the greenhouse effect that occurred between 1984 and 1988.

Whereas the items just discussed deal with future environmental problems, the first Cambridge item asks for respondents' perceptions of how "overall quality of the environment around here" compares with that of 5 years ago. The results reflect a substantial increase in the percentage perceiving a worsening of their local environment from 1983 to 1989 (from 34% to 49%). Given the fact that surveys have consistently found the public more likely to see environmental problems as serious at the state or national levels than at the local level (e.g., Dunlap and Van Liere, 1977), it is striking that far more people see their local environment becoming worse than see it becoming better (in 1990, 31% indicated "better," and the rest volunteered "about the same" or "don't know").

In recent years, one of the most publicized ways in which local environmental quality deteriorates has been through contamination of water supplies, and the next two sets of Cambridge data reflect the importance of this problem to the public. First, between 1981 and 1988 the percentage indicating that they believe "most [underground sources of water] are contaminated with chemicals or other pollutants" or "as many are contaminated as are uncontaminated" nearly doubled, from 28% to 54%. Second, those indicating that the "quality and safety" of their drinking water has worsened over the past 5 years increased by half, from 31% to 46%, in only 5 years.

Although fear of water contamination and toxic contamination in general have led to

Table 5
Recent Trends in the Perceived Seriousness of Environmental Problems

National Survey	Question	Percentage Response by Year (19—)										
		80	81	82	83	84	85	86	87	88	89	90
(1) Roper[a]	Will be a "serious problem" 25 to 50 years from now:											
	(a) severe air pollution	68				70				82		
	(b) severe water pollution	69				71				82		
	(c) shortage of water supplies	57				53				66		
	(d) the "greenhouse effect"	—				37				65		
	(e) overpopulation	52				56				61		
(2) Cambridge[b]	"Overall quality of the environment around here" worse than five years ago				34	33	—	32	32	46	49	55
(3) Cambridge[c]	"Most" or "many" underground sources of water are contaminated with chemicals or other pollutants		28	—	29	37	40	39	50	54		
(4) Cambridge[d]	"Quality and safety of your drinking water" is worse than five years ago						31	31	34	45	45	46
(5) Cambridge[e]	Feel the "greenhouse effect" is a "very" or "somewhat" serious problem			43	—	—	—	63	—	71	75	
(6) Roper[f]	"Environmental pollution" viewed as "very serious threat" to citizens					44	—	—	—	62		
(7) Cambridge[g]	Environment volunteered as one of "the two most important problems" facing the U.S. today			2	—	—	—	—	5	8	16	21

[a]See footnote e of Table 3 for full question.

[b]Full question: "Do you think the overall quality of the environment around here is very much better than it was five years ago, somewhat better than it was five years ago, slightly better than it was five years ago, slightly worse, somewhat worse, or very much worse than it was five years ago?" Results show three "worse" categories combined. Results for 1983–1986 are reported in Cambridge Reports, Inc. (1986, p. 3). More recent results are reported in Cambridge Reports/Research International (1990, p. 4).

[c]Full question: "There are a lot of sources of underground water in the United States. Some people say many of these sources are contaminated with chemicals and other pollutants. I'd like to know how you feel about this. Do you think most underground sources of water are contaminated, as many underground sources are contaminated as are uncontaminated, not very many are contaminated, or none are contaminated?" Results show "most" and "as many are as are not" combined. Results are reported in Americans for the Environment (1989, p. 5-25).

[d]Full question: "Do you think the quality and safety of your drinking water is very much better than it was 5 years ago, somewhat better than it was 5 years ago, slightly better than it was 5 years ago, slightly worse, somewhat worse, or very much worse than it was 5 years ago?" Results are reported in Cambridge Reports/Research International (1990, p. 5).

[e]Full question: "Actually, the greenhouse effect, which is a gradual warming of the Earth's atmosphere, is believed to be caused by carbon dioxide and other gases accumulating in the atmosphere and preventing heat from escaping into space. Some people have expressed concern that the greenhouse effect could lead to harmful changes in ocean levels and weather patterns. Just from this information, do you feel the greenhouse effect is a very serious problem, a somewhat serious problem, not too serious a problem, or not a serious problem at all?" Results are reported in Cambridge Reports, Inc. (Cambridge, MA), *Trends and Forecasts,* October 1989, p. 7.

[f]Full question: "Here is a list of some different things people have said are threats to our society. For each one would you tell me whether you think it is a very serious threat these days to a citizen like yourself, a moderately serious threat, not much of a threat, or no threat at all? First, environmental pollution." Results are reported in Roper's (New York, NY) *The Public Pulse,* Research Supplement, June 1989, p. 1.

[g]Full question: "What do you think are the two most important problems facing the United States today?" Results show combined first- and second-choice responses. Results are reported in Cambridge Reports, Inc., *Trends and Forecasts,* June 1989, p. 5, and September 1989, p. 6, and Cambridge Reports/Research International (1990, p. 3).

growing apprehension among the public about the quality of their local environment, the past few years have also seen an explosion of attention to large-scale environmental problems, ranging from regional problems such as acid rain to global problems such as ozone depletion and global warming. The fifth item taps the last issue, asking respondents about the degree to which they see the greenhouse effect as a problem, and it is clear that there has been a major increase in the proportion of the public viewing this phenomenon as problematic. By 1989, fully three-fourths of the public had come to see the greenhouse effect as at least somewhat serious.

Not only has there been a significant increase in the degree to which the public perceives environmental problems, ranging from local to global, as serious over the past several years, but there has been a concomitant rise in the perception of such problems as real threats to human well-being. This increase can be seen in the responses to the sixth item (although there are only two data points), in which by 1988 Roper found that 62% indicated that environmental pollution was a "very serious threat . . . to a citizen like yourself" (up from 44% in 1984), a response that probably reflects the emergence of hazardous and toxic wastes as major pollutants in the 1980s. Although I am not aware of any comparable items being used in the 1970s, it is hard to imagine that such a large majority of the public would have seen environmental pollution as a "very serious threat" to themselves in the 1970s.

Finally, mirroring the significant rise in the perceived seriousness of environmental problems that has occurred during the 1980s, such problems have begun to re-emerge as leading responses to MIP questions in the last half of the decade. The Cambridge MIP data reported in Table 5 show that in 1990, fully 21% volunteered some type of environmental problem as one of the two most important problems facing the United States today, ranking it fourth behind drugs at 39%, government spending at 26%, and other social problems at 22%. This is the highest percentage of respondents volunteering environmental problems since the April 1970 Harris survey reported in Table 2, and it exceeds the 13% reported by Harris for 1972. This strongly suggests that the environment has re-emerged as a salient issue (even judged by the extremely stringent criterion of showing up on volunteered MIP lists) to the public.[12]

Increasing Threat and Declining Quality

I argued above that there has been an increase in the degree to which environmental problems are perceived as real threats. Because I believe that this represents a major shift from the 1970s, and that it is a prime factor in the significant rise in public concern over environmental quality documented in Tables 4 and 5, I want to focus on the issue in more detail. In 1987 and again in 1989 Cambridge presented lists of environmental problems to samples of the public and asked them to rate the degree of threat posed by each problem—first to "the overall quality of the environment" and then to "personal health and safety"—on a scale of 1 to 7 (1 = no threat at all, 7 = a large threat). Table 6 shows the percentages of "6" and "7" responses (clearly indicating a high degree of perceived threat) for both environmental and personal threats over the 2 years. (The problems are ranked in the order of their rating as "high personal threats" in 1989.)

Several aspects of the results in Table 6 are worth mentioning. First, there was substantial increase (over 10%) in the perceived threat, both to the environment and personally, for nearly all of the problems over the 2-year period. Second, by 1989 majorities of the public (often around two-thirds) were rating most of the problems as

Table 6

Perceived Environmental and Personal Threats from Various Problems, 1987 and 1989

Problem	High Environmental Threat[a]			High Personal Threat[b]		
	1987	1989	Change	1987	1989	Change
(1) Disposal of hazardous waste material	65	71	+6	62	69	+7
(2) Contamination of underground water supplies	52	67	+15	47	65	+18
(3) Air pollution (general)	47	67	+20	—	—	—
(4) Pollution of our rivers, lakes, and oceans	54	67	+13	46	60	+14
(5) Using additives and pesticides in our food supply	49	55	+6	49	60	+11
(6) Depletion of the ozone layer in the atmosphere	42	61	+19	39	58	+19
(7) Air pollution caused by business and industry	—	—	—	37	58	+19
(8) Air pollution caused by cars and trucks	—	—	—	32	52	+20
(9) The greenhouse effect	26	52	+26	20	48	+28
(10) Acid rain	38	53	+15	33	43	+10

Note. Results are reported in Cambridge Reports, Inc. (Cambridge, MA), *Trends and Forecasts,* October 1989, pp. 4–6.

[a]Full question: "Here is a card with a scale from '1' to '7,' where '1' means 'no threat at all' and '7' means 'a large threat.' Now I am going to read you a list of potential threats to the overall quality of the environment. Please use the card to tell me how much you think each problem threatens the overall quality of the environment. The more you think the problem threatens overall environmental quality, the higher the number you would give it." The percentages reported are the combined "6" and "7" responses.

[b]Full question "Now I am going to read you several potential problems facing our society. As I read each one, please use the same card to tell me how much you think each problem threatens your personal health and safety. The more you think a problem threatens your personal health and safety, the higher the number you would give it." The percentages reported are the combined "6" and "7" responses.

large threats to the environment as well as to themselves. Third, there is considerable correspondence between the rating of problems as threats to environmental quality and as threats to personal health and safety, indicating that the two aspects of environmental problems increasingly go hand-in-hand. Finally, the overall ranking of the problems indicates that the threat posed by hazardous wastes, especially the possibility of water contamination, is the leading concern of the public. This finding is in agreement with other recent surveys in which the public is asked to rate the seriousness of a range of environmental problems (e.g., Hart/Teeter, 1990).

Reinforcing the increase in the perceived seriousness of environmental problems and the growing threat attributed to such problems is a sense that the overall quality of the environment is clearly deteriorating. Although I have come across no trend data besides the Cambridge item (which asks about changes in the local environment over the past 5 years) that bear directly on this issue, several recent surveys have found that a substantial segment of the public sees environmental quality as deteriorating. For example, a 1990 NBC/*Wall Street Journal* poll asked, "Overall, do you think the environment in the United States has gotten better, worse, or stayed about the same over the last 20 years?" Although two-thirds felt it had gotten worse, only 16% felt it had gotten better, and 16% that it had stayed about the same (Hart/Teeter, 1990). In a similar vein, a recent Harris (1989) survey that asked, "do you think that in 50 years' time the environment in the world as whole will be much better, a little better, a little worse, or much worse than it is today?" found 81% responding "worse" and only 18% indicating "better."

In short, despite considerable governmental and societal efforts at environmental protection over the past two decades, there is a widespread perception that the quality of the environment—from the local to the global level—is deteriorating. Furthermore, this deterioration is seen as posing a direct threat to the health and well-being of humans. I believe that this growing sense that environmental conditions are becoming truly threatening to our future provides a depth to environmental concern that was largely absent in 1970.

That environmental problems are seen increasingly as threatening, that they are threats to the public at large (rather than to only a segment of the population), and that their effects are ambiguous as well as ominous, are among the conditions that Downs (1972) recognized as possibly forestalling the disappearance of environmental issues from the public's attention. The data reviewed above indicate that Downs was wise to add such qualifications to the issue-attention cycle's applicability to environmental problems.

Summary and Implications

The foregoing evidence on trends in public opinion toward environmental issues strongly indicates that the environmental movement has been extremely successful in attracting and maintaining—for two decades—the public's attention to and endorsement of its cause. We have already seen that large majorities of the public accept environmentalists' definition of ecological problems as serious (increasingly so) and express support for efforts to ameliorate the problematic conditions. In addition, when asked to do so, growing proportions of the public express positive views of the environmental movement and identify with it. Over the past two decades, a number of surveys have asked the public how they feel about the movement, and majorities have indicated that they are at least sympathetic toward it, whereas a significant minority (usually around 10%–15%) claim to be active in the movement (Dunlap, 1989; Mitchell, 1984).

Recent surveys confirm the public's positive image of environmentalism and suggest that after two decades the movement is drawing increasing support from the general population. For example, between 1987 and 1990 Cambridge found that the percentage responding that either they or someone in the household has "donated to or been active in a group or organization working to protect the environment" had risen from 15% to 40% (Cambridge Reports/Research International, 1990). More impressively, 1989 and 1990 Gallup surveys found three-quarters (76% and 73%, respectively) of the public responding "yes" when asked, "Do you consider yourself an environmentalist?" (Gallup and Newport, 1990, p. 7). Thus, it is not surprising that a 1990 survey by Environment Opinion Study (1990) found two-thirds of the public agreeing that "threats to the environment are as serious as the environmental groups say they are" and only one-quarter saying that "environmental groups are exaggerating these threats in order to get the public to pay attention to them." Finally, a 1989 *Business Week*/Harris poll (1989) gave respondents a list of six groups, and for each asked, "Do you think they do more good than harm or more harm than good?" Seventy-five percent indicated that environmentalists did *more good than harm,* second only to their Chamber of Commerce (76%), and well ahead of the third-place consumer advocates. These results reflect an extraordinary level of public participation in, and identification with, the environmental movement and indicate the high degree of credibility and legitimacy that environmentalism has attained within our society.[13]

Of course, as noted in the introduction, the ultimate objective of environmentalists is not simply to convert the public to their cause but to improve the problematic conditions that gave rise to, and continue to drive, the movement. In this regard, the environmental movement has clearly met with considerably less success. This highlights the obvious fact that attaining and maintaining strong and widespread support[14] from the public—and thereby avoiding the fate of passing through Downs's (1972) issue-attention cycle—is not a sufficient condition for a social movement to achieve its goals. Supportive public opinion clearly does not automatically translate into the basic social changes needed for solving major environmental problems.

Yet, public support for environmental protection is extremely high, even higher than two decades ago, and those intent on protecting environmental quality should recognize the tremendous resource that this support offers them. The public is concerned about environmental deterioration and wants to see something done about it. Although it is easy to question the depth of this concern by noting its relatively limited behavioral impact (Dunlap, 1987; Rosenbaum, 1991), perhaps what is needed is better leadership for harnessing public concern and translating it into effective action. Although great emphasis has recently been placed on the necessity of adopting more ecologically sound lifestyles, it is clear that many important changes must be made within political and economic institutions as well as in individual behavior. Thus, environmentalists should continually improve their strategies for channeling the vast reservoir of public support into the political arena (through election campaigns as well as in lobbying efforts) and into the private sector (through direct actions such as economic boycotts as well as through green consuming). Likewise, ecologically aware public officials should realize that they are in a unique position for providing leadership on environmental issues, and the polls suggest that it might be politically astute for them to take the lead in environmental protection. The future of our environment, as well as that of environmentalism, will be heavily influenced by the effectiveness of such leadership.

Notes

1. One can think of Downs's five stages as describing the public's role in the five stages of Mauss's "natural history" model of social-problem movements.

2. Note that the Harris and National Opinion Research Center (NORC) questions focused on perceived levels of pollution in the respondents' vicinity, whereas the Gallup question focused on national priorities. It seems logical that the former would be related more closely to actual pollution levels than would the latter and, because there was very little change in actual pollution levels during this period, the increases found by Harris and NORC are impressive.

3. For a more detailed discussion of conceptual issues concerning public opinion, such as the distinction between salience and the intensity of opinion, see Dunlap (1989).

4. The percentages mentioning the energy shortage also exceeded those mentioning environmental problems in the Harris surveys beginning in 1974 (see Mitchell and Davies, 1978).

5. For a more detailed discussion of this issue, see Dunlap (1989, pp. 124–130).

6. The comparison between the 1970 Gallup poll and the Council on Environmental Quality survey probably exaggerates the degree of decline in environmental concern. First, the 1970 Gallup poll was conducted in late April, immediately after Earth Day. Second, changing societal conditions no doubt influenced the public's selection of priorities from the list. For example, "reducing unemployment" jumped from a tie for seventh to second (25% to 48%) from 1970 to 1980, reflecting the substantial deterioration of the economy over the decade. Third, the results in Table 3 (especially for the National Opinion Research Center spending item), as well as a large amount of cross-sectional data (see, e.g., Mitchell, 1980), indicate the existence of a substantial degree of environmental concern in 1980.

7. A review of the *Gallup Opinion Index* throughout the 1970s reveals that environmental problems reappeared on the "most important problem" list only once after 1973, with 4% in May 1977.

8. For example, two analyses of the state of public concern with environmental quality through the 1970s came to differing conclusions in this regard, with Lake concluding that "it has defied the issue-attention cycle" (1983, p. 232) and Anthony noting that despite its continued strength, "there is also nothing . . . to suggest that Downs' analysis was basically wrong" (1982, p. 19).

9. A good indicator of the degree to which the Carter administration was viewed as pro-environmental is the fact that environmentalists were expressing concern because so many of their leaders were leaving movement organizations for positions within the administration (e.g., Baldwin, 1977).

10. This occurred despite a slight change in the item's wording, from "improving and protecting the environment" to simply "the environment," given to subsamples in the National Opinion Research Center surveys since 1984. Generally, the new wording resulted in a couple of percentage points fewer responding that "too little" is being spent on the environment, but the differences in responses to the two versions are seldom significant. See Dunlap (1989) and Jones and Dunlap (1991) for data on the minimal response differences for the two versions.

11. Unpublished results were kindly made available by The Gallup Organization.

12. Results from recent Gallup polls show a lower percentage naming environment as a "most important problem"—thus far peaking at 8% in April 1990 (see Gallup and Newport, 1990)—but this is with a question that asks for only a single MIP. A 1989 Media General/Associated Press poll that asked what respondents think will be the country's single MIP 10 years from "now" found the environment mentioned by 12%, placing it second only to drugs with 17% (reported in *The Polling Report*, New York, NY, 4 Dec. 1989, p. 1).

13. For additional data on public involvement in and perceptions of the environmental movement, see Dunlap and Scarce (1991).

14. Support for environmental protection has always been widely dispersed among the major segments of society, and there has been minimal change in this regard over the past two decades (see Jones and Dunlap, 1991; Morrison and Dunlap, 1986).

References

Americans for the Environment. 1989. *The rising tide: Public opinion, policy, and politics.* Washington, DC: Author.

Anthony, R. 1982, May. Polls, pollution, and politics: Trends in public opinion on the environment. *Environment,* 24:14–20, 33–34.

Baldwin, D. 1977, June. Environmentalists open the revolving door. *Environmental Action,* 9:13–25.

Business Week/Harris Poll. 1989, 29 May. The public is willing to take business on. *Business Week,* p. 29.

Buttel, F. H. 1975a. Class conflict, environmental conflict, and the environmental movement: The social bases of mass environmental beliefs, 1968–1974. Unpublished Ph.D. dissertation, University of Wisconsin—Madison.

——— . 1975b. The environmental movement: Consensus, conflict, and change. *Journal of Environmental Education,* 7:53–63.

Cambridge Reports, Inc. 1986. Paying for environmental quality. *Bulletin on Consumer Opinion,* No. 112. Cambridge, MA: Author.

——— . 1989. *The rise of the green consumer.* Cambridge, MA: Author.

Cambridge Reports/Research International. 1990. *The green revolution and the changing American consumer.* Cambridge, MA: Author.

Downs, A. 1972. Up and down with ecology—The "issue-attention cycle." *Public Interest,* No. 28:38–50.

Dunlap, R. E. 1987, July/August. Polls, pollution, and politics revisited: Public opinion on the environment in the Reagan era. *Environment,* 29:6–11, 32–37.

——— . 1989. Public opinion and environmental policy. In *Environmental politics and policy,* ed. J. P. Lester, pp. 87–134. Durham, NC: Duke University Press.

Dunlap, R. E., and D. A. Dillman. 1976. Decline in public support for environmental protection: Evidence from a 1970–1974 panel study. *Rural Sociology,* 41:382–390.

Dunlap, R. E., and R. Scarce. 1991. The polls—A report: Environmental problems and protection. *Public Opinion Quarterly,* 55: In press.

Dunlap, R. E., and K. D. Van Liere. 1977. Further evidence of declining public concern with environmental problems: A research note. *Western Sociological Review,* 8:108–112.

Environment Opinion Study, Inc. 1990. *A survey of American voters: Attitudes toward the environment.* Washington, DC: Author.

Erskine, H. 1972. The polls: Pollution and its costs. *Public Opinion Quarterly,* 36:120–135.

Friends of the Earth. 1982. *Ronald Reagan and the environment.* San Francisco: Author.

Funkhouser, G. R. 1973. The issues of the sixties: An exploratory study in the dynamics of public opinion. *Public Opinion Quarterly,* 33:62–75.

Gallup, G., Jr., and F. Newport. 1990, April. Americans strongly in tune with the purpose of Earth Day 1990. *Gallup Poll Monthly,* No. 295, pp. 5–14.

Gillroy, J. M., and R. Y. Shapiro. 1986. The polls: Environmental protection. *Public Opinion Quarterly,* 50:270–279.

Harris, L. 1989, 14 May. Public worried about state of environment today and in future. *The Harris Poll,* No. 21, pp. 1–4.

Hart/Teeter. 1990. NBC News/*Wall Street Journal*: National Survey No. 6. Washington, DC: Author.

Hays, S. P. 1987. *Beauty, health, and permanence: Environmental politics in the United States, 1955–1985.* New York: Cambridge University Press.

Holden, C. 1980. The Reagan years: Environmentalists tremble. *Science,* 210:988–991.

Hornback, K. E. 1974. Orbits of opinion: The role of age in the environmental movement's attentive public, 1968–1972. Unpublished Ph.D. dissertation, Michigan State University.

Jones, R. E., and R. E. Dunlap. 1991, August. The social bases of environmental concern: Have they changed over time? Revision of a paper presented at the 1989 meeting of the Rural Sociological Society, Seattle, WA.

Ladd, E. C. 1982, February/March. Clearing the air: Public opinion and public policy on the environment. *Public Opinion,* 5:16–20.

Lake, L. M. 1983. The environmental mandate: Activists and the electorate. *Political Science Quarterly,* 98:215–233.

Lester, J. P. 1989. Introduction. In *Environmental politics and policy,* ed. J. P. Lester, pp. 1–9. Durham, NC: Duke University Press.

Mauss, A. L. 1975. *Social problems as social movements.* Philadelphia: Lippincott.

McEvoy, J., III. 1972. The American concern with environment. In *Social behavior, natural resources, and the environment,* ed. W. R. Burch, Jr., N. H. Cheek, Jr., and L. Taylor, pp. 214–236. New York: Harper and Row.

Mitchell, R. C. 1980. Public opinion on environmental issues. In *Environmental quality: The eleventh annual report of the Council on Environmental Quality.* Washington, DC: U.S. Government Printing Office.

——— . 1984. Public opinion and environmental politics in the 1970s and 1980s. In *Environmental policy in the 1980s: Reagan's new agenda,* ed. N. J. Vig and M. E. Kraft, pp. 51–74. Washington, DC: Congressional Quarterly Press.

——— . 1990. Public opinion and the green lobby: Poised for the 1990s? In *Environmental policy in the 1990s: Toward a new agenda,* ed. N. J. Vig and M. E. Kraft, pp. 81–99. Washington, DC: Congressional Quarterly Press.

Mitchell, R. C., and J. C. Davies, III. 1978. *The United States environmental movement and its political context: An overview.* Discussion Paper D-32. Washington, DC: Resources for the Future, Inc.

Morrison, D. E. 1986. How and why environmental consciousness has trickled down. In *Distributional conflicts in environmental-resource policy,* ed. A. Schnaiberg, N. Watts, and K. Zimmerman, pp. 187–220. New York: St. Martin's Press.

Morrison, D. E., and R. E. Dunlap. 1986. Environmentalism and elitism: A conceptual and empirical analysis. *Environmental Management,* 10:581–589.

National Opinion Research Center. 1989. *General social surveys, 1972–1989: Cumulative codebook.* Chicago: Author.

Nixon, W. 1991, January/February. 1990—The year of the environment. *E: The Environmental Magazine,* 2:30–37.

Peters, B. G., and B. W. Hogwood. 1985. In search of the issue-attention cycle. *Journal of Politics,* 47:238–253.

Pierce, J. C., K. M. Beatty, and P. R. Hagner. 1982. *The dynamics of American public opinion.* Glenview, IL: Scott, Foresman.

Portney, P. R., ed. 1984. *Natural resources and the environment: The Reagan approach.* Washington, DC: Urban Institute Press.

Rosenbaum, W. A. 1991. *Environmental politics and policy,* 2nd ed. Washington, DC: Congressional Quarterly Press.

Sabatier, P., and D. Mazmanian. 1980. The implementation of public policy: A framework of analysis. *Policy Studies Journal,* 8:538–560.

Schoenfeld, A. C., R. F. Meier, and R. J. Griffin. 1979. Constructing a social problem: The press and the environment. *Social Problems,* 27:38–61.

Smith, T. W. 1985. The polls: American's most important problems: Part I. National and international. *Public Opinion Quarterly,* 49:264–274.

Vig, N. J., and M. E. Kraft, eds. 1984. *Environmental policy in the 1980s: Reagan's new agenda.* Washington, DC: Congressional Quarterly Press.

Index

Abbey, Edward, 57
ACORN, 28
Advocacy science, 22
AFL-CIO, 84
Alabamians for a Clean Environment, 46
Alar, 20, 22, 86
Alaska Public Interest Lands Act, 88n
American Association of Retired Persons, 85
American Civil Liberties Union, 84
American Federation of Scientists, 85
American Lung Association, 84
American Society for International Law, 72
Amicus Journal, The, 55
Anarchism, 56–58
Animal liberation, 7
Antinuclear power movement, 28, 78, 79
Antiwar movement, 28, 39, 56
Appropriate technology, 78
Arctic National Wildlife Refuge, 88n
Audubon Society (*see* National Audubon Society)

Bearing witness, 59
Bhopal, 44–45, 63, 68, 74n, 107
Big outside, The, 58
Bioregionalism, 7, 54, 78
Blueprint for the environment, 80
Brower, David, 14, 17, 19
Browning-Ferris Industries, 44
Brundtland, Madame Gro Harlem, 74
Bureau of Land Management, 87
Bureaucratization, of environmental movement, 3, 5, 11, 24, 52, 55, 79
Burger King, 61n
Bush, George, 59–60, 72, 103, 106

Carson, Rachel, 2, 14, 19, 28, 52, 65
Carter, Jimmy, 4, 22, 29, 69, 81, 102, 114n
Cato Institute, 79
Charismatic leadership, decline of, 21–23
Chemical Waste Management, 41, 45
Chernobyl, 63, 66, 68, 74n, 107
Citizen's Clearinghouse for Hazardous Waste, 18(t), 19, 29, 30, 35, 42
Citizens Environmental Coalition, 30
Citizens for a Better Environment, 31
Civil rights, 6, 28, 39, 40, 43, 45, 46, 56, 85
Clamshell Alliance, 78
Clean Air Act, 80, 88n
Clean Air Coalition, 84
Clean Water Act, 80, 88n
Closing circle, The, 28
Commission for Racial Justice, 30, 41, 43
Commoner, Barry, 19, 28, 78
Community right-to-know, 28, 31, 34
Compromise, by environmentalists, 24, 33, 52, 55, 57, 77, 79
Concerned Citizens of South Central Los Angeles, 40
Conflict resolution, 40, 44, 45–47
Conservation biology, 61
Conservation directory, 25n
Conservation Foundation, 14, 24n, 79
Conservation International, 17, 18(t), 71
Conservation movement, 1–2, 8n, 12, 14, 17, 21, 24, 51, 56, 79, 82
Conservation organizations, 5, 12, 14, 15 19, 21, 31, 32, 51, 91
Conservation/preservation schism, 2, 8, 51, 79
Consumer Federation of America, 84
Convention on the Control of Transboundary Movements of Hazardous Wastes, 69
Convention on the Law of the Sea, 74
Co-optation, 3, 4
Corfu incident, 64
Council of All Beings, 56
Council on Environmental Quality, 3, 101, 102
Countermovements, 4, 51, 52, 60, 61n

Note: n means note and (t) means table.

Dallas Alliance Environmental Task Force, 46
Darman, Richard, 72
Debt-for-nature swaps, 71
Deep ecology, 6, 7, 51–62, 78
Defenders of Wildlife, 12, 13(t), 15, 16, 22, 85
Demise of social movements, 1, 3, 5, 8
Department of Interior Secretary:
  Lujan, Manuel, 59–60, 72
  Watt, James, 4, 15, 17, 82, 88n, 102, 103, 105, 106
Development of Third World, 68, 71–74
Diet for a new America, 54
Dinosaur National Monument, 2
Direct action, 6, 17, 18(t), 19, 20, 29, 42, 55, 56, 61n, 79, 87
  (*See also* Monkey-wrenching; Tree-spiking)
Direct mail, 11, 12, 15–17, 19, 23, 24n, 82–84
Diversity of the environmental movement, 1, 5–8, 19, 59, 72
Dow Chemical, 33
Downs, Anthony, 90–91, 96, 97, 101–102, 106, 112, 113, 114n
Dust bowl, 2

Earth Day:
  first, in 1970, 2, 7, 8, 12, 14–16, 19, 20, 28, 64, 78, 82, 89, 90, 92, 94, 96, 97, 101, 102
  twentieth, in 1990, 1, 3, 8, 15, 43, 59, 87, 89, 90, 107
Earth First!, 6, 7, 17, 18(t), 20, 51, 52, 56–60, 61n, 79
Earth Island Institute, 18(t), 19
Ecodefense, 57
Eco-feminism, 7, 54, 61
ECONET, 58
Ecosophy, 53
Ecotage (*see* Direct action; Monkey-wrenching; Tree-spiking)
Ecotage!, 25n
Eco-terrorists, 7, 57
Education, as tactic, 12, 17, 19, 28, 30, 57, 59, 87
Educational, Scientific, and Cultural Organization, United Nations (UNESCO), 65
Elitist image of environmental movement, 6
Emergency Conservation Committee, 85
Endangered Species Act, 59
Energy crisis, 96, 99n, 100–102, 114n

Environmental Action, 13(t), 14, 16–17, 20, 25n, 40, 78, 82, 85
Environmental agenda for the future, An, 14, 80
Environmental Defense Fund, 8n, 13(t), 14, 22, 23, 28, 32, 33, 42, 56, 79, 80, 85
Environmental Ethics, 61
Environmental impact statements, 3, 20, 65, 66
Environmental justice, 6, 30, 31, 34–35, 39–49, 55, 58, 61, 85
Environmental Liaison Centre, 67
Environmental Policy Institute, 13(t), 14, 78
Environmental problems:
  difference from conservation issues, 2, 14, 15, 19, 21, 32, 51, 91
  unique nature of, 5, 11, 14, 16, 21, 80, 107
Environmental Protection Agency, 3, 28, 29, 41, 72, 74n, 81, 83, 86, 87, 96, 102, 103, 105
Environmental Protection Agency Director:
  Gorsuch, Anne, 4, 102, 103, 106
  Reilly, William, 72
Environmental racism, 6, 30, 40–41
Environmental Task Force, 78
Environmental wars, The, 60
Exxon Valdez, 15, 63, 68, 74n

Fair share, 28
Federal Bureau of Investigation, 60
Federal Environmental Pesticide Act, 68
Feminism (*see* Women and the environment)
Food and Drug Administration, 87
Ford Foundation, 14
Ford, Gerald, 81
Foreman, Dave, 56–58
Fox, the, 25n
Fragmentation of movements, 3, 5, 7
Friends of the Earth, 13(t), 14–17, 19, 68, 78, 85

Gandhi, 54
Gibbs, Lois, 19, 29, 30, 33, 34, 42
Glen Canyon Dam, 57
Global Tomorrow Coalition, 73
Globalization of environmental movement, 6–7, 63–76
Globescope, 73
Gorsuch, Anne, 4, 102, 103, 106
Grand Canyon, 2, 91
Grass Roots Environmental Organization, 30
Grassroots organizing, 6, 7, 17, 24, 27–37, 39–49, 51, 54, 55, 80, 82, 84, 86
  success of, 33–35

Great Depression, 2
Green consumerism, 7, 54, 59, 77, 86–87, 113
Greenpeace, 6, 7, 13n(t), 17, 18(t), 19, 20, 51, 58, 60, 68, 69, 79
Greenpeace Action, 13n(t)
Green pledge, 87
Green rage, 56
Greens, 54–55, 57, 59, 73, 78
Group of 10, 14, 47, 84
Growth rates of environmental organizations, 3-5, 11, 12, 15–16, 19, 22, 24, 51, 66, 77, 79–80, 83, 85, 91, 106
Gulf Coast Tenants Organization, 40

Hayes, Denis, 8
Health, as environmental issue, 2, 6, 7, 22, 27–35, 51, 52, 67, 70, 80, 89, 110, 111n, 112
Helsinki Declaration on the Protection of the Ozone Layer, 70
Heritage Foundation, 79
Holmes, Bill, 60
Humane Society of the United States, 85

Informative science, 19
Institutionalization of environmental movement, 3-5, 21, 51, 106
Interest groups, 3-5, 24, 25n, 91
International Atomic Energy Agency, 66
International Biological Programme, 65
International Council of Scientific Unions, 65
International Environmental Law Centre, 75n
International Geophysical Year, 65
International perspectives on the study of climate and society, 70–71
International Union for Conservation of Nature and Natural Resources, 65
International Whaling Commission, 71
Issue-attention cycle, 90–91, 96, 97, 101–102, 106, 112, 113, 114n
Izaak Walton League, 12, 13(t), 16

Jackson, Reverend Jesse, 43
Jackson, Senator, 91
Jim Crow, 40
Job blackmail, 40
Johnson, Lyndon, 65, 91

Krupp, Fred, 79

Labor, 23, 40, 43
League of Conservation Voters, 14, 17, 18(t), 19, 20, 24n, 82
Locally unwanted land uses (LULUs), 6, 40
Love Canal, 6, 19, 27, 29, 30, 34, 42
Lowery, Reverend Joseph, 43
Lujan, Manuel, 59-60, 72

Media, 2-5, 16, 35, 67, 68, 72, 83, 89-91, 96, 97, 101, 105-107
Mendez, Chico, 60
Minorities, 6, 7, 27, 30, 32, 34, 39–49, 85
Minority Peoples' Council, 45, 46
Monkey-wrench gang, The, 57
Monkey-wrenching, 6, 17, 56–58 (*See also* Direct action; Tree-spiking)
Montreal Protocol on Substances That Deplete the Ozone Layer, 70
Moon landing, 67
Mothers of East Los Angeles, 30, 40
Muir, John, 2, 8, 51, 79
Muskie, Edmund, 91

Nader, Ralph, 84
Naess, Arne, 52–53, 55, 61
National Audubon Society, 2, 11, 12, 13(t), 16, 21, 30, 32, 51, 55, 84, 85, 88n
National Council of Churches, 85
National Environmental Policy Act, 65, 66, 92
National Parks and Conservation Association, 12, 13(t), 16, 25n, 85
National Rainbow Coalition, 43
National Taxpayers Union, 84
National Toxics Campaign, 17, 18(t), 19, 29, 30, 35, 42
National Wildlife Federation, 11, 12, 13(t), 23, 25n, 32, 42, 51, 80, 84, 85
Native Americans for a Clean Environment, 40
Natural history of social movements, 3–4, 90–91, 96, 106, 114n
Natural Resources Council of America, 84
Natural Resources Defense Council, 13(t), 14, 20, 22, 23, 28, 32, 33, 42, 51, 80, 84–86, 88n
Nature Conservancy, 17, 18(t), 19, 68, 79, 84, 85
New York Coalition for Alternatives to Pesticides, 30
NGO Environmental Forum, 66–67
NIABY, not in anyone's back yard, 27, 35
NIMBY, not in my back yard, 6, 7, 27, 35, 47

Nixon, Richard, 81
Nuclear power movement, 28, 78, 79

Office of Management and Budget Director Richard Darman, 72
Office of Technology Assessment, 65
Organization for Economic Cooperation and Development, 69
OSHA, 84, 87
Our common future, 74

Partial Nuclear Test Ban Treaty, 66
People Concerned about MIC, 45, 46
People First!, 60
Pinchot, Gifford, 2, 8, 79
Planet of the Year, 59, 107
Political action committees, 20, 82
Population, 2, 51, 53, 72–74, 80, 107, 108(t)
Preservation/conservation schism, 2, 8, 8n, 51, 79
Preservation movement, 2, 8n
Professionalization of environmental movement, 5, 11, 21–23, 33, 51, 55, 79, 83
Proposition 13, 102
Proposition 65, 34, 87
Public opinion/support, 1, 3, 4, 7, 11, 12, 15, 16, 22, 24, 28, 35, 65–69, 77, 80, 81, 86, 89–116

Radical environmentalism, 6, 7, 24, 51–62, 77–79, 85–87
Rainforest Action Network, 6, 18(t), 51, 58–59, 61n, 68, 79
Rainforest Alliance, 18(t)
Rainforest Information Center, 58–59
Reagan, Ronald, 4–7, 8n, 15, 17, 20, 35, 55, 64, 69, 73, 78, 81, 84, 102, 103, 105, 106
Redwood Summer, 59, 60
Reformist environmentalism, 6, 51, 52, 54–56, 60, 78, 79, 85, 86
Reilly, William, 72
Resource mobilization, 8n
Resources for the Future, 79
Right-to-know, of community, 28, 31, 34
Rollins Environmental Services, 41, 45, 46
Ronald Reagan and the American environment, 103
Roosevelt, Franklin, 2
Roosevelt, Theodore, 2
RSR Corp., 44, 46

Sandoz, 63, 74n
Santa Barbara oil spill, 2
Science:
as environmentalist tactic, 5, 12, 14, 21–23, 29, 30, 33, 55, 68, 87, 90, 106
as impetus for environmental movement, 2, 5, 19, 28, 65, 67, 70, 82
grassroots distrust of, 27, 31–32
Science and survival, 19
Scientific Committee on Problems of the Environment, 65
Sea Shepherd Society, 6, 17, 18(t), 20, 51, 58, 79
Seed, John, 56, 58
Sierra Club, 2, 7, 11, 12, 13(t), 14, 15, 19, 21, 23, 30, 32, 42, 51, 55, 56, 60, 68, 77, 80, 82–85, 88n, 91
Sierra Club Legal Defense Fund, 18(t), 19, 23, 84
Silent spring, 2, 14, 19, 28, 52, 65
Silicon Valley Toxics Coalition, 30
Social ecology, 7, 55, 58, 78
Social justice (*see* Environmental justice)
Social movements, natural history of, 3–4, 90–91, 96, 106, 114n
Society of Friends, 59
Southern Christian Leadership Conference, 43
Southwest Organizing Project, 40
Southwest Research and Information Center, 31
Spaceship Earth, 67
Specialization of environmental organizations, 6, 23, 83
Standing, 20, 21, 64
Steelworker's Union, 84
Success of environmental movement, 1, 8, 23, 55, 89, 113
Sununu, White House Chief of Staff, 72
Superfund, 28, 34, 45, 80, 87, 88n, 103

Tax laws, 12, 21, 22
Technical Assistance Grant Program, 34
Tennessee Valley Authority, 2
Texans United, 30
Texas Department of Health, 47
Toxic Avengers of Brooklyn, 40
Toxic Substances Control Act, 68
Toxic wastes and race, 41
Toxics in Your Community Coalition, 30
Trail smelter, 64
Treaty on Principles Governing the Activities of States in the Exploration

and Use of Outer Space, Including the Moon and Other Celestial Bodies, 66
Tree-spiking, 7, 17, 60
(*See also* Direct action; Monkey-wrenching)
Trumpeter, The, 53, 61
Trust for Public Lands, 85

Union Carbide, 44–45
Union of Concerned Scientists, 85
United Church of Christ, 30, 43
United Nations, 65, 69, 72
United Nations Commission on Environment and Development, 71
United Nations Conference on Environment and Development, 1992, 74
United Nations Conference on the Human Environment, Stockholm, 64–67, 72
United Nations Educational, Scientific, and Cultural Organization (UNESCO), 65
United Nations Environmental Programme, 67
United Nations Law of the Sea Treaty, 72
U.S. Forest Service, 2, 86, 87
U.S. National Research Council, 70
Urban Environment Conference, 40

Valdez principles, 85
Vegetarianism, 54
Vietnam, 56
Voluntary simplicity, 54, 78

Watson, Paul, 58
Watt, James, 4, 15, 17, 82, 88n, 102, 103, 105, 106
West Harlem Environmental Action, 40
Wilderness Act, 88n
Wilderness movement, 2
Wilderness Society, 12, 13(t), 15, 21, 22, 51, 55, 56, 84, 85, 88n
Women and the environment, 7, 27, 29, 56–58
Women of All Red Nations, 30
Work on Waste, 30
World Bank, 59, 71
World Climate Conference, 70
World Commission on Environment and Development, 74
World Meteorological Association, 65
World Peace Through Law, 72
World Resources Institute, 79
World War I, 2, 12, 13(t)
World War II, 2, 3, 12, 13(t), 28
World Wildlife Fund, 17, 18(t), 19, 65, 68, 79, 85

About the Editors

Riley E. Dunlap is Professor of Sociology and Rural Sociology at Washington State University, and past chair of the American Sociological Association's Section on Environmental Sociology, the Rural Sociological Society's Natural Resources Research Group, and the Society for the Study of Social Problems' Environmental Problems Division. He has conducted wide-ranging research on environmental attitudes and activism over the past two decades and is currently working with the George H. Gallup International Institute on an international environmental opinion survey.

Angela G. Mertig is a graduate of Ripon College and a Washington State University Ph.D. candidate in sociology specializing in environmental sociology, human ecology, and social movements.